MONITORING FOR
HEALTH HAZARDS AT W

MONITORING FOR HEALTH HAZARDS AT WORK

BY FRANK S GILL AND
INDIRA ASHTON

FOREWORD BY
RICHARD WARBURTON
Director General, RoSPA

△ RoSPA

THE ROYAL SOCIETY FOR THE
PREVENTION OF ACCIDENTS

GRANT MCINTYRE
MEDICAL & SCIENTIFIC

Copyright © 1982 by
Grant McIntyre Ltd
90–91 Great Russell Street
London WC1B 3PY

All rights reserved. No part of this
publication may be reproduced, stored
in a retrieval system, or transmitted,
in any form or by any means,
electronic, mechanical, photocopying,
recording or otherwise
without the prior permission of
the copyright owner

First published 1982

Photoset by Enset Ltd
Midsomer Norton, Bath, Avon
and printed and bound
in Great Britain by
Butler & Tanner Ltd
Frome, Somerset

DISTRIBUTORS

Throughout the world except
North America and Australia
   Blackwell Scientific Publications Ltd
   Osney Mead, Oxford, OX2 0EL

USA
   Blackwell Mosby Book Distributors
   11830 Westline Industrial Drive
   St Louis, Missouri 63141

Canada
   Blackwell Mosby Book Distributors
   120 Melford Drive, Scarborough
   Ontario, M1B 2X4

Australia
   Blackwell Scientific Book Distributors
   214 Berkeley Street, Carlton
   Victoria 3053

British Library
Cataloguing in Publication Data

Gill, Frank S.
   Monitoring for health
   hazards at work.
   I. Title II. Ashton, Indira
   363.1′1   RC967

ISBN 0–86286–029–6

CONTENTS

FOREWORD, ix

ACKNOWLEDGEMENTS, viii

PREFACE, x

UNITS AND ABBREVIATIONS, xii

1 DUST, 1
  Introduction, 1
  Equipment required for filtration sampling, 2
    Filters, 3
    Filter holders, 3
    Sampling pumps, 5
  Direct reading instruments, 7
  Calibration of a rotameter using a soap bubble method, 12
  The measurement of total airborne dust using a 25mm or 37mm open face filter, 15
  The measurement of airborne respirable dust using a cyclone separator, 19
  The sampling and counting of airborne asbestos fibres, 22
  The choice of filter and filter holder to suit a specific dust, fume or mist, 26
  To trace the behaviour of a dust cloud using a Tyndall beam, 27
  Further reading on dust, 28

2 GASES AND VAPOURS, 29
  Introduction, 29
  Equipment available, 30
    Collection devices, 30
      Containers, 30
      Adsorption methods, 31
      Passive samplers, 32
    Pumps, 33
    Tube holders, 34
    Adsorbent tubes, 35
    Colourimetric detector tubes, 35
    Direct reading instruments, 37
    General, 37
  To obtain a personal sample for solvent vapours using an adsorbent tube, 37
  The collection of gases using a sampling bag, 41
  The sampling for gases using a bubbler, 43
  To measure the airborne concentration of a gas using a colourimetric detector tube, 45
  Further reading on gases and vapours, 51

Contents

3 HEAT, 52
Introduction, 52
Equipment available, 53
  Dry bulb thermometers, 53
  Wet bulb thermometers, 53
  Sling psychrometer, 53
  Aspirated psychrometer, 54
  Digital humidity meter, 54
  Continuous recording of temperature and humidity, 55
  Globe thermometer, 57
  Kata thermometer, 57
  Integrating instruments, 57
  The psychrometric chart, 59
  Heat indices, 60
The measurement of the thermal environment, 62
Use of the Kata thermometer chart, 66
Use of the globe thermometer chart, 69
Further reading on heat, 69

4 VENTILATION, 70
Introduction, 70
Equipment available, 71
  Pressure measuring instruments, 71
  Air velocity measuring instruments, 74
  Calibration devices, 79
  Barometric pressure instruments, 79
The measurement of airflow in ducts, 79
The measurement of pressure in ventilation systems, 84
To measure the performance of a suction inlet, 86
The measurement of natural air infiltration rate in a room, 89
Calibration of an anemometer in an open jet wind tunnel, 91
Further reading on ventilation, 93

5 NOISE, 94
Introduction, 94
Equipment available, 98
  Sound level meters, 98
  Noise dosemeters, 98
  Calibration, 99
To measure a steady workroom noise, 100
The measurement of the spectrum of a continuous noise by octave band analysis, 102
To measure the L_{eq} of a fluctuating workroom noise, 105
The use of a personal noise dosemeter, 106
Further reading on noise, 108

6 LIGHT, 109
Introduction, 109
Units used in lighting, 109
Equipment available, 111
  Photometers, 111
  Hagner Universal photometer, 112
  Daylight factor meter, 112
To measure the daylight factors in a room, 113
To undertake a lighting survey of a workroom, 116
Further reading on lighting, 121

Contents

7 OTHER HAZARDS, 122
   Introduction, 122
   Ionising radiation, 122
      Instruments available, 123
   Non-ionising radiation, 124
   Microbiological hazards, 126
   Further reading on radiation and microbiological hazards, 126

8 SURVEYS, 128
   Introduction, 128
   Planning, 128
   Labour and assistance, 129
   Results, 129
   Further reading on surveys and standards, 130
   Equipment check lists, 131
   Suppliers of equipment, 142
   Addresses, 144
   List of occupational hygiene consultants, 147
   Analytical services, 148
   Engineering control consultant and contractors, 148

INDEX, 149

ACKNOWLEDGEMENTS

The authors would like to thank the following who have helped in some way in the preparation of this book: E.C.Ashton, T.E.Ashton, D.Blunt, the British Standards Institution, The Chartered Institute of Building Services, J.Conyers, J.C.Edwards, J.H.Edwards, J.Fish, G.Franklyn, A.D.F.Gill, R.M.Gill, J.M.Harrington, I.J.King, K.Knight, J.Longmore, C.McDonald, MRC Pneumoconiosis Unit, N.Peacock, S.Richards, A.Sisman, N.Torrance, H.A.Waldron, C.Williams, A.J.Woolley and the many companies who have provided photographs and advice on their products.

FOREWORD

Whilst occupational hygiene and safety is not perhaps thought of as an exact science, it increasingly demands more detailed and accurate measurement of conditions at the workplace. This is so because of the need to justify the cost of preventive effort and to allay concern that changes in working practice or the introduction of new substances or materials does not conflict with expectations for a safe and healthy working environment. If there is to be full compliance with legal standards and, as important, those standards are to be maintained, the co-operation and understanding of managers and workpeople is of major importance.

This book is of particular value as it fills, within one cover, a real need for practical guidance on the range of instruments available and the significance of what they record. I hope it will encourage everyone with an interest in health in the workplace to make the fullest possible use of instruments so that they can contribute personally towards the raising of standards in their undertaking. A book like this goes a long way to overcoming the natural resistance of many to what seems, on the surface, to be a highly technological subject best left to experts.

RICHARD WARBURTON
Director General, RoSPA

PREFACE

As a result of the Health and Safety at Work etc. Act of 1974 workplace environments and their related health hazards have come under closer scrutiny. Duties of employers under Section 2 of that Act require workplaces and the working environment to be 'safe and without risks to health. . . .' The Safety Representatives Regulations made under the same Act has brought a greater awareness to shop floor workers and others that their place of work may be the cause of ill health and that their employers have a duty to safeguard health and ensure their comfort.

Over the short period of time since the enactment of that piece of legislation it has been necessary for management, often prompted by the workforce, to enquire into the possible health hazards present in the workplaces in their charge. Even places traditionally thought of as safe, such as offices, have experienced hazards associated with the new technology found there, and in manufacturing and service industries processes use chemicals and materials that are new and often inadequately tested for possible health risks. With the increased use of machinery comes an increase in unacceptable noise levels and radiation of heat and electromagnetic waves of various wavelengths and energies whilst more detailed and intricate work requires better illumination. In order to assess the degree of risk to the workforce it is often necessary to measure the concentration or intensity of the suspected hazard, which may be airborne dust, gas, vapour, heat, noise or some other pollutant; and it may also be necessary to check the performance of environmental control devices such as ventilation and lighting systems.

Although there is a body of people trained in the skills of workplace monitoring and control for health and comfort who practice under the name of 'occupational or industrial hygienists', at the time of writing there are less than 100 registered as Operational or Professional in Great Britain. Many of these work for large companies or government bodies. Therefore the burden of finding someone competent to monitor a workplace falls on the shoulders of management who search within their companies for people with scientific training and experience to undertake this somewhat onerous task involving the use of scientific instruments and equipment which require skill and insight to operate satisfactorily. People who often find themselves in this position are: company chemists, safety officers and advisers, nurses, works engineers and others.

The range of instruments and equipment available in the field is

Preface

bewildering and guidance on selection and use is required. Many of the monitoring and analysis techniques require the skills of analytical chemists, acousticians, ventilation engineers, microbiologists and health physicists but the workplace sampling and simple monitoring must of necessity, be done by less qualified 'in house' staff.

This book is offered as a guide to those who are keen to know how to handle the types of equipment available and who are required as part of their duties to monitor the workplace for health. Also is is hoped that trade union safety representatives and shop stewards will consult it to assist them to understand and, in some cases, advise management on methods of workplace measurement and sampling.

The book is written around a series of instruction sheets covering a variety of sampling and monitoring procedures grouped in suitable chapters. By way of introduction to the sheets a review of the types of instruments and equipment to be used is made and some of the basic principles of the techniques are described. Check lists are offered to assist in the assembly of items for a survey on a particular topic and, where appropriate, a method of recording the results is recommended. Lists of suppliers of equipment are given together with their addresses current at the time of going to press.

However, it is important to realise that there is no short cut to a full understanding of the science and behaviour of workplace pollution. The techniques described provide only an indication of the degree of hazard that is present. It must also be understood that workplace pollution of whatever kind does not occur evenly distributed in time or space therefore one single reading or measurement will not represent the workplace as a whole or of that place tested over the working shift or any other period of time. Therefore if major decisions are to be made based upon workplace environmental measurement then the best possible professional advice must be sought to plan and execute detailed surveys and from the results to provide sound judgement on their meaning and to suggest the best course of action. A list of consultant occupational hygienists is given who are professionally qualified to provide a full range of surveys and advice.

Readers are warned that some workplace environments are contaminated with inflammable gases, vapours and dusts and that electrical equipment within them is required to be flameproof. This requirement extends to items of monitoring or sampling equipment both permanent and portable. Instruments which have been passed as suitable for use in such environments are issued with a BASEEFA certificate and care must be taken to ensure that only these are used. In case of doubt the manufacturer should be consulted.

UNITS AND ABBREVIATIONS

The more common units used in workplace environmental measurement

Unit	Dimension	SI	Imperial	Conversion
Length	L	m mm	ft in	ft×0.3048 = m in×25.4 = mm
Area	L^2	m² mm²	ft² in²	ft²×9.29×10⁻² = m² in²×645.2 = mm²
Volume	L^3	m³ (1000 litre) 1 (litre)	ft³ gallon	ft³×2.832×10⁻² = m³ gallon×4.546 = litre
Mass	M	kg g (gramme) mg	lb oz grain	lb×0.4536 = kg oz×28.35 = g gr×64.79 = mg
Airborne Concn. of substance (Mass)	$\dfrac{M}{L^3}$	mg m⁻³	grain ft⁻³	gr ft⁻³×2288 = mg m⁻³
(Volume) (Particle)	— —	Parts per million mp cm⁻³ mp ft⁻³		(millions of particles per cm³ (ft³))
Acceleration	$\dfrac{L}{T^2}$	m s⁻² gravity = 9.81m s⁻²	ft sec⁻²	ft s⁻²×0.305 = m s⁻²
Density	$\dfrac{M}{L^3}$	kg m⁻³ (g/l)	lb ft⁻³	lb ft⁻³×16.02 = kg m⁻³
Flow rate (Mass)	$\dfrac{M}{T}$	kg s⁻¹	lb hr⁻¹	lb hr⁻¹×1.26×10⁻⁴ = kg s⁻¹
(Volume)	$\dfrac{L^3}{T}$	m³ s⁻¹	ft⁻³ min⁻¹ gall hr⁻¹	ft³ min⁻¹×4.719×10⁻⁴ = m³ s⁻¹ gall hr⁻¹×1.263×10⁻⁶ = m³ s⁻¹
Force	$\dfrac{ML}{T^2}$	N (Newton) (N = kg.m s⁻²)	lb_f	lb_f×4.448 = N
Energy Heat Quantity	$\dfrac{ML^2}{T^2}$	J (Joule) = Ws = Nm kW hour	Btu	Btu×1055 = J kW hour×3.6×10⁶ = J kilocalorie×4187 = J
Heat Flow Power	$\dfrac{ML^2}{T^3}$	W	H.P. Btu hr⁻¹	H.P.×745.7 = W Btu hr⁻¹×0.291 = W
Latent Heat	$\dfrac{L^2}{T^2}$	kJ kg⁻¹	Btu lb⁻¹	Btu lb⁻¹×2.326 = kJ kg⁻¹
Specific Heat	$\dfrac{L}{T^2 \text{ temp}}$	kJ kg⁻¹°C	Btu lb⁻¹°F	Btu lb⁻¹°F×4.187 = kJ kg⁻¹°C

Unit	Dimension	SI	Imperial	Conversion
Pressure	$\dfrac{M}{T^2L}$	Pa (Pascal) = N m^{-2} bar ($\times 10^5$ = Pa)	lb$_f$ ft^{-2} lb$_f$ in^{-2} inches water (4°C) inches mercury (0°C)	lb ft^{-2} $\times$ 47.88 = Pa lb in^{-2} $\times$ 6895 = Pa in H$_2$O $\times$ 249.1 = Pa in Hg $\times$ 3386 = Pa
Torque	$\dfrac{ML^2}{T^2}$	Nm	lb$_f$ ft	lb$_f$ ft $\times$ 1.356 = Nm
Velocity	$\dfrac{L}{T}$	m s^{-1}	ft min^{-1} ft sec^{-1}	ft m^{-1} $\times$ 5.08 $\times$ 10^{-3} = m s^{-1} ft s^{-1} $\times$ 0.305 = m s^{-1}
Viscosity (Dynamic)	$\dfrac{M}{TL}$	Pa s (Ns m^{-2}) Poise (dyne sec cm^{-2})	lb.s ft^{-1}	lb.s ft^{-1} $\times$ 47.88 = Pa s Poise $\times$ 0.1 = Pa s
(Kinematic)	$\dfrac{L^2}{T}$	m^2 s^{-1} Stokes (cm^2 s^{-1})	ft^2 s^{-1} in^2 s^{-1}	ft^2 s^{-1} $\times$ 9.29 $\times$ 10^{-2} = m^2 s^{-1} in^2 s^{-1} $\times$ 6.452 $\times$ 10^{-4} = m^2 s^{-1} stokes $\times$ 10^{-4} = m^2 s^{-1}
Luminous intensity		candela (cd)	candle (int)	candle $\times$ 0.981 = cd
Luminous flux		lumen (lm) (lm = lcd sr)		
Illuminance		lux (lx = lm m^{-2})	foot candle lumen ft^{-2}	ft candle $\times$ 0.1076 = lx lm ft^{-2} $\times$ 0.1076 = lx
Luminance		cd m^{-2}	foot lambert candela in^{-2}	ft lambert $\times$ 3.426 = cd m^{-2} cd in^{-2} $\times$ 1550 = cd m^{-2}

Some useful initials

OSHA	Occupational Safety and Health Administration (Act) (USA)
NIOSH	National Institute of Occupational Safety and Health (USA)
ANSI	American National Standards Institute
ACGIH	American Conference of Government Industrial Hygienists
HASAWA	Health and Safety at Work Etc. Act. (UK)
HSC	Health and Safety Commission (UK)
HSE	Health and Safety Executive (UK)
EMAS	Employment Medical Advisory Service (UK)
BOHS	British Occupational Hygiene Society (UK)
IOH	Institute of Occupational Hygienists (UK)
BSI	British Standards Institution (UK)
MRC	Medical Research Council (UK)
PSPS	Pesticides Safety Precautions Scheme (UK)
WHO	World Health Organisation
ILO	International Labour Office
BASEEFA	British Approvals Service for Electrical Equipment in Flammable Atmospheres

Units and Abbreviations

Abbreviations used in the text

bar	bar	N	Newton or number of air changes per hour
Btu	British thermal unit	oz	ounce
C	concentration	p	pressure
°C	degree Celsius or centigrade	Pa	Pascal
cd	candela	ppm	parts per million
cm	centimetre	R	reflectance
D	diameter	s	second
d	diameter or density correction factor	sr	steradian
dB	decibel	t	time or temperature °C
E	illuminance	T	absolute temperature Kelvin
f	number of fibres	V	volume or volume flow
ft	foot	v	velocity
g	gramme	W	watt
g	acceleration due to gravity	μg	microgram
gall	gallon	μm	micrometer
gr	grain	θ	Kata thermometer range
h	hour	Φ	luminous flux
Hg	mercury	ρ	density
Hz	Herz (cycles per second)	Δ	delta meaning 'change of'
I	luminous intensity		
in	inch		
J	Joule		
L	luminance or level		
l	litre		
lb	pound		
lb$_f$	pound force		
lm	lumen		
lx	lux		
m	metre		
mb	millibar		
mC	milliiCurie		
mg	milligram		
min	minute		
mp cm^{-3}	millions of particles per cubic centimeter		
mp ft^{-3}	millions of particles per cubic foot		

Multiples of SI units

Name	Symbol	Factor
tera	T	10^{12}
giga	G	10^{9}
mega	M	10^{6}
kilo	k	10^{3}
hecto	h	10^{2}
deca	da	10^{1}
deci	d	10^{-1}
centi	c	10^{-2}
milli	m	10^{-3}
micro	μ	10^{-6}
nano	n	10^{-9}
pico	p	10^{-12}
femto	f	10^{-15}
atto	a	10^{-18}

CHAPTER 1
DUST

Introduction

Airborne dust is ubiquitous and workplaces are no exception; every operation and action releases into the air a certain amount of dust. Movement of people can release dust from clothing and skin. Even dust that has settled on floors and flat surfaces is made airborne by air currents as people move about their work. Windborne dust enters buildings particularly in dry weather and more so in densely populated areas. Add to that particles released by the operations within a workplace—handling of materials, machining, cutting, drilling, grinding, milling, sanding and planing of items being manufactured—and a dusty working atmosphere can be produced. Fortunately most dust is harmless but in sufficient concentrations it can cause discomfort and unpleasantness. At such levels it is termed a 'nuisance dust'. However, some dusts are distinctly harmful, giving rise to carcinoma, chronic lung disease, asthma, bronchitis and other disorders.

Not only does the chemical composition of the material and its airborne concentration determine its detrimental effects but also the particle size influences the part of the lungs where the material is deposited. Large particles are collected in the nose and throat whilst smaller ones are deposited further into the lung, the next stage of collection being in the upper airways, i.e. the bronchi and bronchioles, where self-clearing action by the ciliary movement takes place. The very small particles reach the deepest parts, the alveoli, where the oxygen transfer takes place. Some particles are breathed out again and some are removed by body fluids but others remain and cause physical and chemical reactions which can be harmful in both the short and long term, sometimes leading to permanent lung damage.

Dust size is measured in the unit called a micro-metre (μm) which is a 1000th part of a millimetre; a human hair is above 30μm in diameter. Unfortunately most particles of dust are irregular in shape and rarely spherical or circular therefore it is difficult to quote size. The important feature is how dust behaves when it is airborne, particularly how rapidly it settles in still air. In the field of occupational health the term 'aerodynamic diameter' is used to denote the particles' size. This is the diameter of a theoretical spherical particle of unit density (1g ml^{-1}) which settles at the same speed as the particle in question; thus any irregularly shaped particle can be assigned an aerodynamic diameter. Dust with an aerodynamic diameter of 7μm and above will not normally reach the alveoli but particles below that size will and are therefore termed 'respirable'.

Dust from coal, sand and most hard rocks is harmful at respirable size but is normally cleared from the lung if larger; whereas pollens, spores and mists are larger and can cause problems in the upper respiratory passages. Fumes from molten metal are very small, below $1\mu m$, and can give rise to metal fume fever and more serious disorders. When determining the concentration of airborne dust it is important to understand what size of dust is to be measured as this influences the method of sampling.

There are two basic methods of sampling airborne dust. The first, and most common, is to draw a known volume of air through a pre-weighed filtering device by means of an air pump, weighing the filtering device after to determine the mass of dust collected. By dividing that mass by the total volume of air drawn through, an average dust concentration is obtained for the sampling period. The second method involves using an instrument which gives a direct reading of the dust concentration at any instant of time but may or may not give an average over a period of time. Filtration systems are available which are lightweight enough for workers to wear to determine their personal exposure to the airborne dust. If they move about in and out of dust clouds or if the emission varies in concentration the readings still monitor the average exposure.

The filtration method can also be used to monitor a working area using one static position throughout the sampling period. In this case the equipment should, if possible, be attached to a tripod or fixed object in the area under study. In order to achieve sufficient accuracy it is important to weigh the filters to 0.01mg. This is beyond the precision of most laboratory balances. Also the flow rates of the air pumps must be checked with a calibrated rotameter so that the total air flow rate passing through the filter is accurately known. A section on calibrating a rotameter is included in this chapter.

The direct reading instruments are more bulky and are unsuited to personal monitoring; they are used to measure a working area rather than an individual. A review of some of these instruments is given later.

Equipment required for filtration sampling

An assembly of items is required to make up a sampling 'train' consisting of:
1   a filter,
2   a filter holder,
3   a suction pump,
4   some connecting tubing,
5   a harness.

In addition a rotameter is required to check air flow rates. These items are discussed in more detail below.

Filters

These are made of a variety of materials with properties suited to different types of analysis. This is covered in more detail in the section entitled 'The choice of filter and filter holder to suit a specific dust, fume or mist' found later in this chapter. Filters are mainly 'fibrous' in structure, made from glass, paper, polystyrene or from a 'membrane' of cellulose derivatives, PVC and polycarbonate. There is also a sintered silver filter available. The correct filter must be chosen to suit the airborne contaminant to be sampled and the subsequent analysis to be undertaken; for example, some can be dissolved in chemicals for further analysis of the collected dust, some can be made transparent for optical microscopic examination of the material, whilst others allow the collected dust to remain on the surface for scanning electron microscopic examination. Some filters are more sensitive to atmospheric moisture content than others and need to be pre-conditioned before weighing.

Filters used for dust sampling are available in diameters of 13, 25, 37, 47, 50, 55mm and larger. The smaller filters are more commonly used for personal sampling and the larger ones for high volume work. Pore sizes vary between $0.1\mu m$ to $10\mu m$ but it should be noted that the pore size does not limit the size of the dust to be collected, that is, a $5\mu m$ pore size filter is capable of capturing dusts smaller than $5\mu m$ by virtue of the inertial and electrostatic forces that occur within the filter medium. In fact it is often desirable to use pore sizes in the range $5-10\mu m$ even for respirable dusts, to reduce the loading on the sampling pump. For particle and fibre counting with a light microscope it is useful to have the filter marked with a squared grid as it assists in focussing.

Filter holders

In order to accommodate the filter and make a connection via tubing to the pump, a holder is used which can be hung from a harness or attached to a static position. Filter holders for personal sampling are usually 25 or 37mm in diameter and can be either open face or partially covered as illustrated in figs. 1.1, 1.2 and 1.3. It is important to select the correct type of holder to suit the material to be collected. Details of which type to use are given in table 1.5.

Where size selection is required, as with sampling for respirable dust, a cyclone-type holder is used which imparts a centrifugal force to the particle as it passes through. By this means the larger particles are separated from the smaller which are collected on the filter. The respirable dust is separated from the total by arranging the airflow rate to be correct for the design of the cyclone. Illustrated in fig. 1.4 is a cyclone which holds a 37mm diameter filter but adapters are available to allow 25mm filters to be used. This cyclone required an airflow rate of 1.9 $l\,min^{-1}$.

For static sampling for respirable dusts some units have a combined

Chapter 1

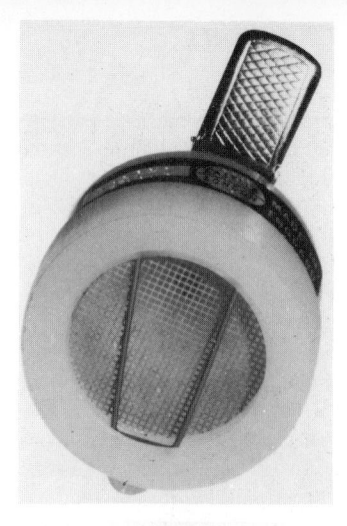

Fig. 1.1. 37mm dia open face filter holder (Casella Ltd).

Fig. 1.2. UKAEA filter holder (Casella Ltd).

Fig. 1.3. Modified UKAEA filter holder (Casella Ltd).

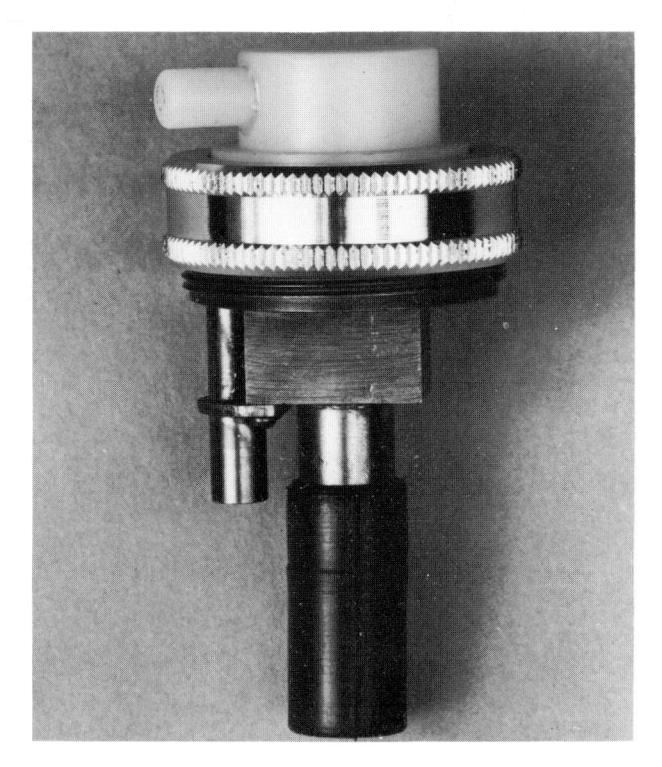

Fig. 1.4. Cyclone filter holder (Rotheroe and Mitchell Ltd).

pump and filter holder, for example the MRE 113A shown in fig. 1.8. This is commonly used in mines in Great Britain. In this, sampler size selection for the respirable particles is achieved by means of an 'elutriator' where the larger sizes settle on a series of parallel plates between which the dust-laden air passes before reaching the 55mm diameter filter. As with all size selection the sampled air flow rate must be fixed and constant.

Sampling pumps

There are three basic types of pump unit: the dry vane rotary as used in the Rotheroe and Mitchell pumps; the single acting diaphragm used in the Casella, Dupont and MSA pumps; and the double acting piston used in the Pitman pumps. Table 1.1 summarises the characteristics of each type. From a dust sampling point of view it is important to note that the rotary pumps produce the smoothest flow rates; the piston and diaphragm pumps produce a pulsating flow and require a flow 'smoother' to

Chapter 1 be added to the sampling train if particle size selection is being made. Some of the latest pumps have a flow smoother integral with the pump.

Table 1.1. Characteristics of personal sampling pumps

Pump type	Diaphragm	Piston	Rotary
Power consumption	Low	Medium	High
Battery size	Small	Medium	Large
Weight	Low	Medium	High
Repair	Simple	Difficult	Moderate
Cost	Cheap	High	Medium
Flow smoothness	Strongly pulsating one pulse per rev.	Mildly pulsating two pulses per rev.	Smooth, three or four pulses per rev.
Pressure drop limits	4.9 kPa	None	None
Valve problems*	Can leak	Can leak	None (no valves)
Manufacturer	Casella, DuPont, MSA	Pitman	Rotheroe & Mitchell

*It is advisable not to run a pump without a filter connected to it as particles of dust can be drawn on to the valve seats or into the rotor.

Fig. 1.5. Personal sampling pump (Rotheroe and Mitchell Ltd).

Dust

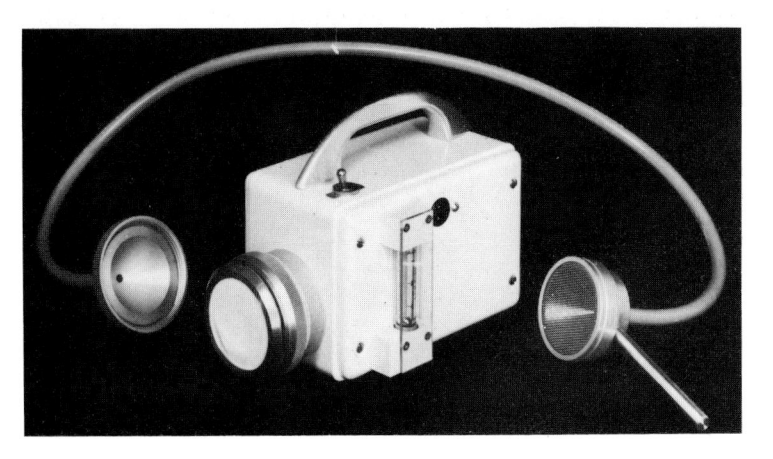

Fig. 1.6. High volume sampling pump (Rotheroe and Mitchell Ltd).

To collect quantities of dust in a reasonable period of time that are adequate for subsequent weighing or other forms of analysis, fairly high flow rates may be required. This is normally easier to achieve for static sampling where mains powered pumps capable of up to 100 l min^{-1} are available.

For personal sampling, easily portable battery operated pumps are used some of which can achieve up to 4.5 l min^{-1}. Some have rotameter-type flow meters built in so that flow rates can be visibly checked from time to time, but diaphragm pumps are normally fitted with a stroke counter to indicate the number of pulses of the diaphragm that have occurred during the sampling period and from this the total flow can be calculated. These medium flow pumps are all battery powered using rechargeable batteries and the suppliers include chargers in their catalogues. The latest types have automatic constant flow control.

Direct reading instruments

Compared with the filter method which requires calculations and the use of accurate weighing techniques, direct reading instruments have the advantage that localised peaks of concentration can be identified and remedial measures immediately put into effect. They can be coupled to a recording system to obtain time-weighted average concentrations. Unfortunately, due to their bulk it is almost impossible to obtain personal exposure measurements with them. Also they are far more expensive to buy than the equipment required for the filtration methods.

For their operation the instruments rely upon one of the following physical principles: the scattering of light by airborne particles of dust;

Chapter 1

Fig. 1.7. Sampling train being worn on a harness (Rotheroe and Mitchell Ltd).

the beta-ray absorption of a deposit of dust on a mylar film; or the oscillation frequency variation of a crystal of quartz when laden with dust (this latter technique is known as a 'piezo-electric' microbalance). Each of the instruments has a digital display or a meter which gives a reading of dust concentration in mg m^{-3} except one, the Royco, which provides a particle count. Because of the variations of physical properties of dust the instruments cannot be accurate for all types so they require careful

Fig. 1.8. MRE 113A Static respirable dust sampler (Casella Ltd).

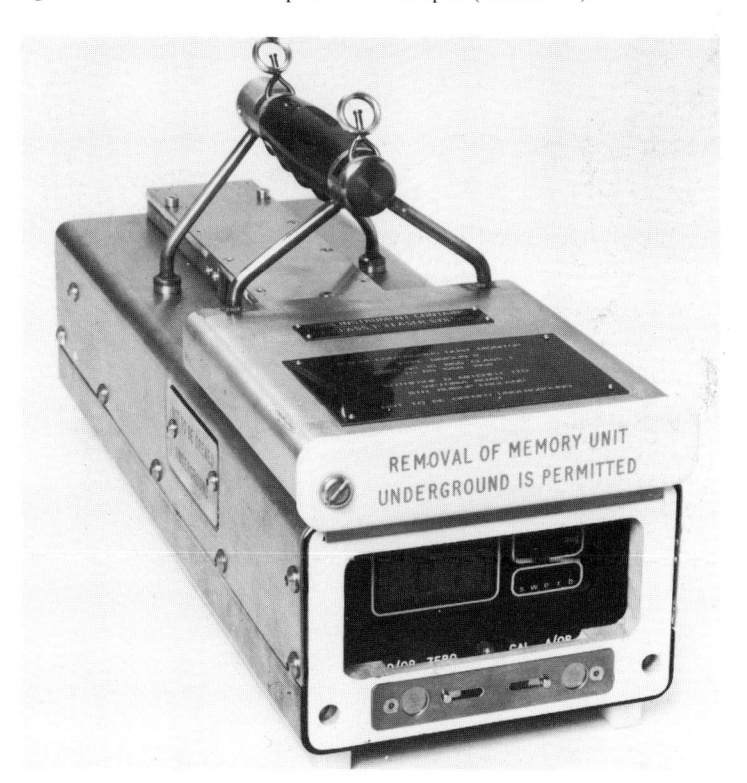

Fig. 1.9. SIMSLIN II Direct reading dust sampler (Rotheroe and Mitchell Ltd).

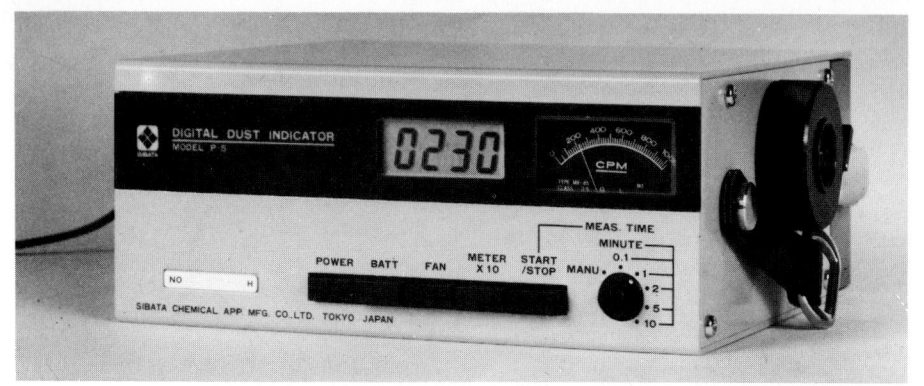

Fig. 1.10. SIBATA P5 Direct reading dust sampler (Rotheroe and Mitchell Ltd).

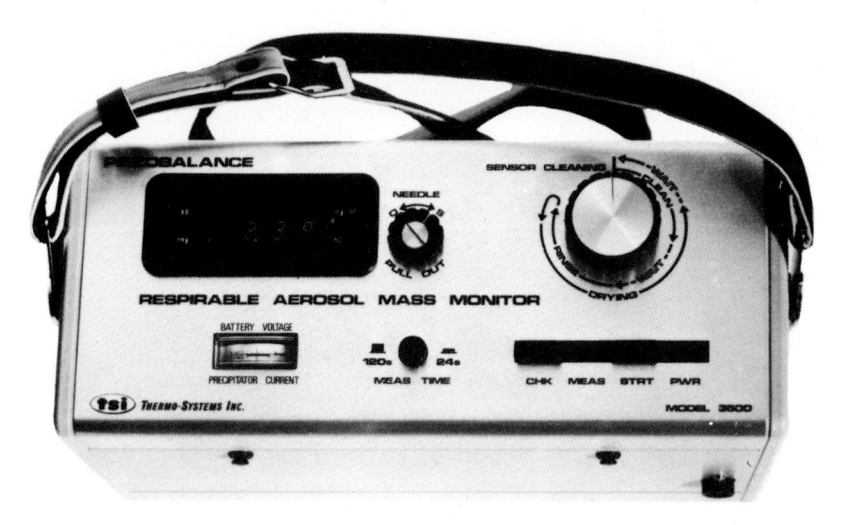

Fig. 1.11. TSI Respirable mass monitor (Bristol Industrial and Research Associated Ltd).

calibration with the dust to be measured in a dust cloud of known concentration. One make, the SIMSLIN, incorporates a facility for collecting a sample of the dust that has passed through the instrument on a membrane filter so that calibration can be achieved by comparing the average recorded dust concentration with that calculated from the weight gain of the filter. Table 1.2 shows some of the direct reading instruments available and gives their salient features.

Calibration of a rotameter using a soap bubble method

Dust

Aim

When sampling for dust or gases in a work situation, from time to time it is necessary to check the flow-rate of air being drawn through the sampling train by the pump. A small portable rotameter is often used

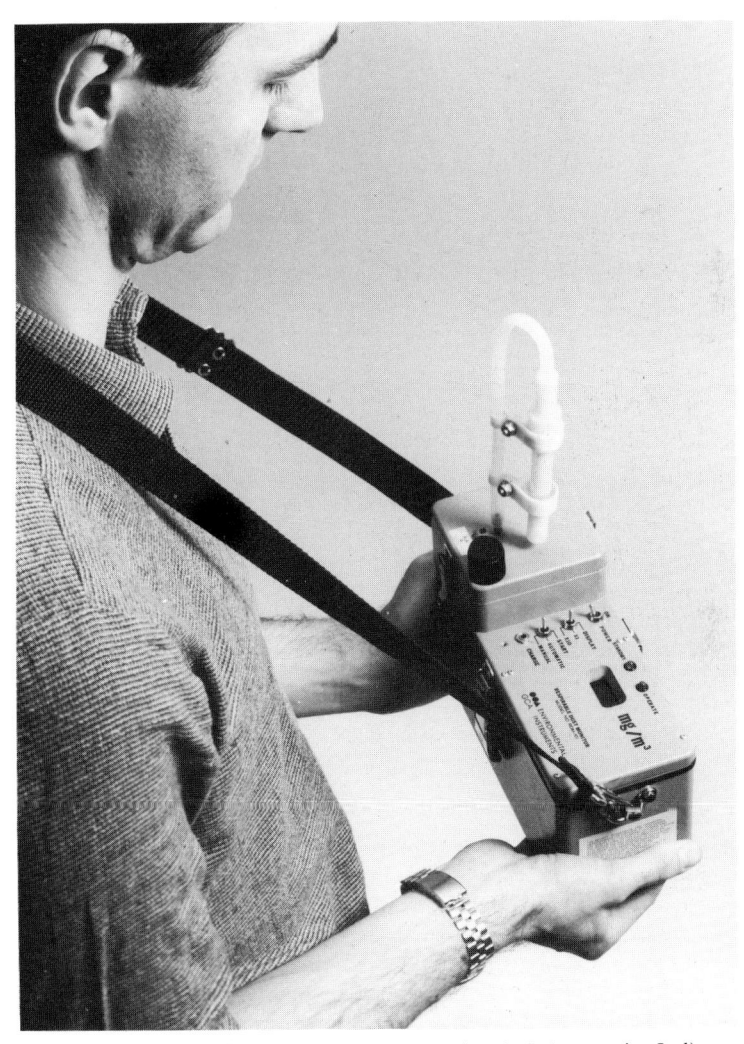

Fig. 1.12. RDM 101 Direct reading dust sampler (Analysis Automation Ltd).

Table 1.2. Details of some direct reading dust sampling instruments

Name and type	Name of UK manufacturer or supplier	Principle of operation	Type of display	Special features
SIMSLIN II	Rotheroe and Mitchell	Light scattering	Digital mg m^{-3}	For respirable size dust, digital recorder and play back facility, filter calibration, battery powered.
Sibata P5	Rotheroe and Mitchell or MDA Scientific (UK) Ltd	Light scattering	Digital mg m^{-3}	Recorder available, automatic timer for regular spot readings, battery powered.
Royco 218	Gelman Sciences Ltd.	Light scattering	Digital (particle count)	Size selective, sampling duration: 1 minute, 10 minute or manual, battery powered, gives particle count not mass concentration.
Royco 220	Gelman Sciences Ltd	Light scattering	Digital or Analogue (particle count)	Simultaneous automatic counting of multiple size ranges, mains powered, size selective, gives particle count not mass concentration, can be used for airborne fibre counts.
RAM I	Analysis Automation Ltd	Light scattering	Digital mg m^{-3}	Total or respirable size selection, recorder available, battery powered.
FAM I	Analysis Automation Ltd	Light scattering	Digital (fibre count)	Designed for fibre counting, length selective, battery or mains powered, digital printer available.
RDM Series: 101, 201, 301	Analysis Automation Ltd	Beta-attenuation	Digital mg m^{-3}	For respirable size dust, battery powered, automatic operation.
TSI Respirable Aerosol Mass Monitor	Bristol Industrial and Research Associates Ltd	Piezo-electric	Digital mg m^{-3}	For respirable size dust, battery powered, vertical elutriator available for cotton dust.

which must first be calibrated. The soap bubble technique is a cheap and reliable method for doing this.

Equipment required

A steady flow-rate sampling pump capable of supplying air at the flow rate of the rotameter to be calibrated, a glass burette graduated in ml, the rotameter to be calibrated, a rubber bulb, a glass T-piece, some flexible tubing of suitable size to connect the items, two stands, boss heads and clamps, tubing clips, liquid soap, cotton wool, stop watch or timer, beaker. The size of the burette will depend upon the range of the rotameter to be calibrated; for medium flow rate pumps that is, 1.0–4.5 l min^{-1} a burette of 250ml capacity is most suitable but for lower flow rates a 100ml size is sufficient.

Some pump manufacturers supply their own soap bubble calibrators which, although more expensive than assembling the equipment above, are self-contained and may be less trouble.

Method

1 Wash the burette with water and then wet the inside surfaces with a thin film of the liquid soap. This can be done by pouring in a little of the soap and washing it down with water from a beaker.
2 Assemble the apparatus as shown in fig. 1.13 and place a small plug of cotton wool at the top of the burette to prevent soap from being drawn into the pump.
3 Place some of the liquid soap in the rubber bulb.
4 Clamp the rotameter in a vertical position using the retort stand and fittings.
5 Start the pump to draw air through the system.
6 Squeeze the rubber bulb gently to release some soap into the air stream so that a bubble will be formed.
7 Time the bubble passing between the 0 and 250ml marks on the

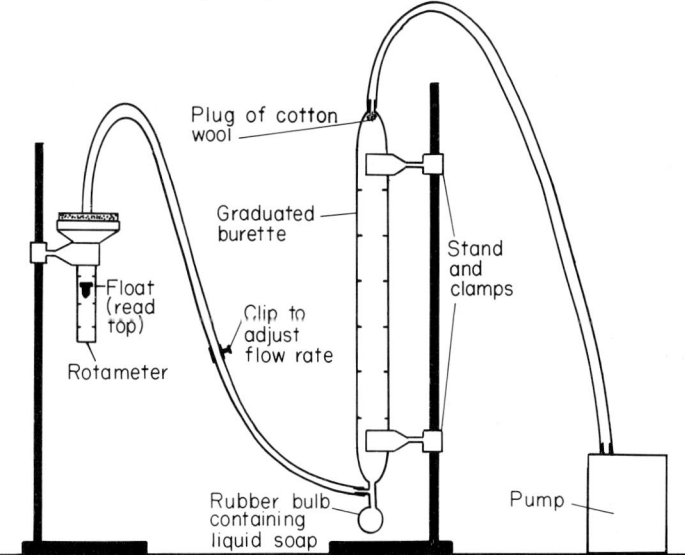

Fig. 1.13. Apparatus to calibrate a rotameter.

burette and repeat this five times for one particular setting of the flow. When observing the position of the bubble for timing it is important to keep one's eye level with the mark.

8 By means of a tubing clamp or by altering the pump flow rate the position of the float in the rotameter can be altered and steps 6 and 7 can be repeated. This should be done for several flow rates throughout the range of the rotameter.

Results and calculations

Using a chart similar to table 1.3 plot the results as they are taken.

To calculate actual flow, F_a, use the expression:

$$F_a = \frac{V}{t} \times 60 \text{ l min}^{-1}$$

where: V is the swept volume of the bubble in litres (1 litre = 1000ml)
and t is the mean time in seconds

Table 1.3. Results sheet for the calibration of a rotameter

Indicated flow on rotameter l min^{-1}		0.5	1.0	1.5	2.0	2.5	3.0	3.5	4.0
Time taken for bubble to travel between marks (sec)	readings: 1								
	2								
	3								
	4								
	5								
Mean of the five readings									
Actual flow l min^{-1}									

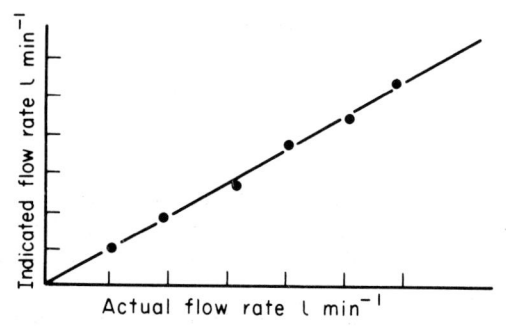

Fig. 1.14. Typical rotameter calibration chart.

Plot the calculated mean flow rates against the rates marked on the stem of the rotameter as shown in the graph in fig. 1.14. This graph should always be carried with the rotameter for checking the pump flow rates during surveys or tests.

Possible problems

1 The rotameter float may be pulsating due to the pump having a pulsating flow characteristic. To reduce this, either fit a flow-smoother into the line or change to a rotating-vane pump.

2 The soap bubble may burst before reaching the end of its travel, due to the sides or the burette being too dry. The burette should be removed from its clamps and the sides wetted with liquid soap and water as in step 1 in the method section.

The measurement of total airborne dust using a 25mm or 37mm open face filter

Aim

This, the most basic of airborne dust measuring techniques, is used for obtaining a general total dust concentration in the breathing zone of a worker or in a particular static situation. The technique is modified if a specific size range or type of dust is required to be collected and assessed. Such modifications involve changing the type of filter and/or filter holder as indicated in later sections of this chapter.

Equipment required

For each place to be simultaneously sampled, a sampling pump capable of a flow rate of 1.0–$4.5\,l\,min^{-1}$, a clean open face filter holder capable of holding a 25mm or 37mm diameter filter, one metre of 7mm internal diameter plastic tubing, a calibrated rotameter (see above), 25mm or 37mm glass fibre filters, forceps, harness, a balance capable of weighing to 0.01mg, labels, Petri-slides and, if static samples are to be collected, then some form of substantial stand such as a heavy photographic tripod and some means of attaching the holder and pump to it such as adhesive tape.

Method

1 Because the weight of filters tends to vary with the humidity of the air, it is wise to allow them to become pre-conditioned by placing each one in a Petri-slide to stand overnight in the room in which they are to be weighed.

2 Weigh the filters to 0.01mg. Always weigh more than required for sampling so that one is used as a control and the others are spares in case

Chapter 1

of accidental damage or contamination. If a filter is accidentally dropped or touched after weighing then it must be discarded as the weight will be altered.

3 Transport each of the weighed filters to the measuring site in a separate labelled Petri-slide.

4 Unscrew the front of the filter holder and using forceps or tweezers carefully place one weighed filter on the grid and replace the front. Do not over-tighten as the filters can easily be damaged. It is wise to number and label each filter holder.

5 Make up the sampling train by connecting the filter holder to the pump by means of the plastic tubing and place in the sampling position. If the measuring position is on a worker then it is best that the whole assembly is held in position by means of a harness (see fig. 1.7).

6 Attach the filter holder face forwards as close to the worker's face as possible on the front of the shoulder strap and hang the pump on the belt. If a static measuring position is to be taken, then attach the filter holder securely to the place to be measured, preferably on a tripod.

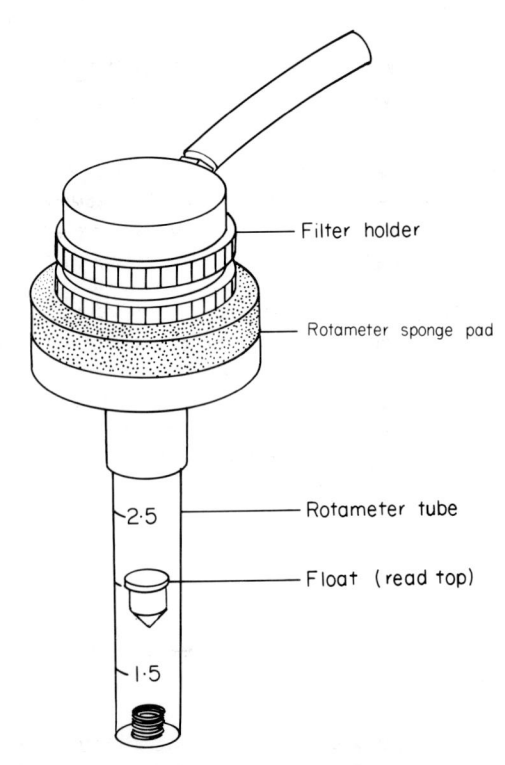

Fig. 1.15. Rotameter in use with open face filter holder (Casella Ltd).

7 Switch on the pump noting the time, and by means of the calibrated rotameter note the flow rate of air passing through the sampling train and adjust if necessary to the required flow rate. To check the flow rate place the face of the filter holder tightly down on to the sponge seal of the rotameter and by holding the glass tube of the rotameter in a vertical position at eye level observe the position of the float as shown in fig. 1.15. The mark level with the top of the float indicates the flow rate. It will be necessary to check the pump flow rate from time to time during the sampling period and adjust if necessary to attempt to maintain a constant flow rate. Some pump flow rates can be adjusted by turning a controlling screw by means of a small screwdriver, others cannot be adjusted. The latest design of pump has an automatic flow adjuster which maintains a constant flow, equal to the setting at the start, throughout the period. Some pumps have a pulse counter or a timer whose value must be noted at the beginning and end of the period. The length of time of sampling will depend upon circumstances, but precision will be lost if an insufficient volume of air is drawn through the filter. Recommended minimum sampling times should be calculated from the following expression:

$$\text{minimum vol. (m}^3\text{)} = \frac{10 \times \text{sensitivity of the balance (mg)}}{\text{suitable hygiene standard (mg m}^{-3}\text{)}}$$

The suitable hygiene standard chosen will depend upon the dustiness of the place sampled but it is suggested that if this is in doubt then one tenth of the TLV should be used. Should it be necessary to stop and restart the pump at any time during the test then all times must be noted and flow rates checked.

8 At the end of the sampling period stop the pump and note the time.
9 After exposure carefully remove the filter using forceps and return it to the labelled Petri-slide.
10 A period of pre-conditioning in the balance room similar to that used for the initial weighing should elapse before re-weighing. The control filter should also be pre-conditioned and re-weighed.
11 Record all readings and results as they are made on a table similar to table 1.4.

Calculations

Determine the total volume of air that has passed through the filter using the expression:

$$\text{Volume of air } V = \frac{\text{flow rate of pump in l min}^{-1} \times \text{duration in min}}{1000} \text{ m}^3.$$

Chapter 1

Table 1.4. Suggested layout for a dust sampling results sheet

Survey .... Sampled by ....

Location .... Date ....

Filter no.	Pump no.	Sampling position	Sampling time			Flow rates		Total volume	Weight gain	Dust conc.	Remarks
			On	Off	Total min	Measured $l\,min^{-1}$	Corr. avg. $l\,min^{-1}$	m^3	mg	$mg\,m^{-3}$	

Note: if the flow rate has changed during the period this calculation must be done for each change of rate and all the volumes added together.

To calculate the true weight gain of the filter proceed as follows:

weight of filter before exposure $= x_1$ (mg)
weight of filter after exposure $= x_2$ (mg)
weight of control filter before $= z_1$ (mg)
weight of control filter after $= z_2$ (mg)
weight of dust on filter $= x_2 - x_1 - (z_2 - z_1)$ (mg)

concentration of dust in air in mg m^{-3} = $\dfrac{\text{wt of dust on filter (mg)}}{V}$

Possible problems

1   Care must be taken to ensure that the filters are not contaminated either accidentally or deliberately by extraneous dust being allowed to come into contact with them. They must always be handled with forceps or tweezers. The UKAEA type of filter holder should be used if there is a risk that large particles of dust, which are unlikely to be inhaled, could be projected towards the filter and should always be used when sampling for heavy dusts such as lead or compounds of lead or mercury.
2   If the pump flow rate has changed unnoticed during the period, then an estimate can be made of the total flow based upon what flows have been noted but the resulting calculations will only be a rough estimate of the true dust concentration.
3   Damaged filters must be discarded and that sample lost.
4   Filters can be overloaded if the sampling rate is too high or if the dust concentration is very dense. If this is suspected a lower flow rate or a shorter sampling period should be adopted.
5   Overloaded filters can lose dust due to handling and transport.

The measurement of airborne respirable dust using a cyclone separator

Aim

In order to separate the respirable fraction of airborne dust from the total dust a modification of the basic open face filter method is adopted. The filter holder used is a cyclone separator which at a flow rate of 1.9 litre min^{-1} will separate dust according to a characteristic curve which approximates to the separation curve of the upper respiratory passages of the body. The technique is similar to the open face filter method except that it is important to maintain a smooth flow rate at a steady 1.9 l min^{-1} throughout the sampling period. For this purpose it is useful to make use of one of the controlled flow rate pumps.

Equipment required

For each place to be sampled simultaneously, a cyclone separator and a

Chapter 1

pump capable of a flow rate of 1.9 l min^{-1}. Pumps which operate on a rotary principle (see table 1.1) provide a sufficiently steady flow rate for satisfactory separation but pumps with a reciprocating action require a flow smoothing device to ensure a smooth flow at the cyclone. The Higgins or (BCIRA) cyclone takes 25mm diameter filters but the SIMPEDS cyclones take 37mm. The remaining equipment required is the same as for the measurement of total dust using an open face filter.

Method

1 Filters should be weighed in the same way as steps 1 and 2 on page 15.
2 Loading the filters into the cyclone varies with the type of separator used. With the Higgins cyclone it is necessary to dismantle the instrument and place the 25mm filter on the grid (see fig. 1.16) and carefully reassemble ensuring that the two halves of the cyclone are not over-tightened to prevent the filter from becoming damaged. With the SIMPEDS cyclone the 37mm filter is assembled into a cassette held together by a wide elastic band (see fig. 1.17). This is best done in a jig provided for the purpose. Adapters are available to enable a 25mm filter to be placed in a 37mm cyclone.
3 The sampling procedure is the same as for the open face filter, steps 5 to 11, except that the sampling rate must be maintained at 1.9 l min^{-1} which should be checked regularly during the sampling period using a calibrated rotameter. The base of the Higgins cyclone can be held tightly against the sponge pad of the rotameter and the tube held vertically at eye level to read. With the SIMPEDS cyclone it is necessary to connect the rotameter to the inlet nozzle by means of rubber tubing.

Calculations

These are exactly the same as with the open face filter method outlined on page 19.

Possible problems

1 If the pump flow rate has changed from 1.9 l min^{-1} during the sampling period then the filter has not collected a representative sample of respirable dust, although a tolerance of 0.1 l min^{-1} above or below that can be accepted. If the flow rate has reduced then dust larger than respirable size will have been collected giving an exaggeratedly high result. The opposite would have occurred if the flow rate had been above 1.9 l min^{-1}.
2 As the mass of a particle of dust varies with the cube of the diameter, and as respirable dust is so small it will be very light in weight. Thus it is often difficult to obtain a sufficiently large weight gain on the filter to give

Fig. 1.16. Exploded view of the Higgins cyclone.

Fig. 1.17. Exploded view of the SIMPEDS cyclone (Casella Ltd).

a reliable result. To minimise this effect sampling durations should be as long as possible.

3 Damaged filters must be discarded as it is impossible to obtain a meaningful weight gain from them.

The sampling and counting of airborne asbestos fibres

Aim

Asbestos is present in many work situations, for example, as thermal

Chapter 1

insulation on pipes, building surfaces, vehicle linings; as an acoustic absorbent material on walls, ceilings, silencers; as a building material such as roofs, walls, pipes and gutters; in vehicle brake linings; and in many other places where its useful properties are utilised. As the health hazard of inhaling fibres from the manufacture, use, fabrication and removal of this material has been clearly demonstrated and because legislation in Great Britain requires the monitoring and control of airborne fibres, it is necessary to establish the airborne concentrations. The procedure for sampling is similar to the open face filter method outlined on pages 15–19.

Equipment required

As for the open face filter method except that the filter shall be one which can be cleared to permit the transmission of light for microscopic examination; 25mm dia cellulose acetate membranes are suitable but it is preferable to have a type with a gridded surface to assist in focussing and dividing up the surface for counting; a means for clearing the membrane (glycerol triacetate (triacetin) is the most common solution but other recently developed methods include acetone vapour, Euparol and DMF); a micro-pipette, a microscope capable of approximately 500× magnification fitted with phase contrast lighting and an eyepiece graticule of the Becket and Walton type (fig. 1.18) or failing that the BS3625 type (fig. 1.19), microscope slides and cover glasses, two push button digital counters.

Method

The proceedure for sampling is the same as with the open face filter as outlined on pages 15–19 except that it is unnecessary to weigh the membrane filter before or after use. Sample volume is important in that sufficient sample needs to be taken to ensure accuracy whilst too much sample can make subsequent analysis difficult or impossible. Only experience can determine the most appropriate sample volume but as a guide, where the expected concentration is in the order of 10 fibres per millilitre ($f\,ml^{-1}$), a ten litre sample is recommended; for $2\,f\,ml^{-1}$, 50 litres and for lower concentrations sample volumes should be in excess of 100 litre.

The sampling period will depend upon the characteristics of the operation and the sample volume to be collected. The Health and Safety Executive use ten minute or four hour periods to assure compliance with the Asbestos regulations. The membrane containing the sample is prepared for microscopic examination as follows:

1 Place a blue paper circle (from the membrane box used to separate the filters) on to a white tissue. This is to act as a marker.
2 Label a cleaned glass microscope slide at one end with the sample

Dust

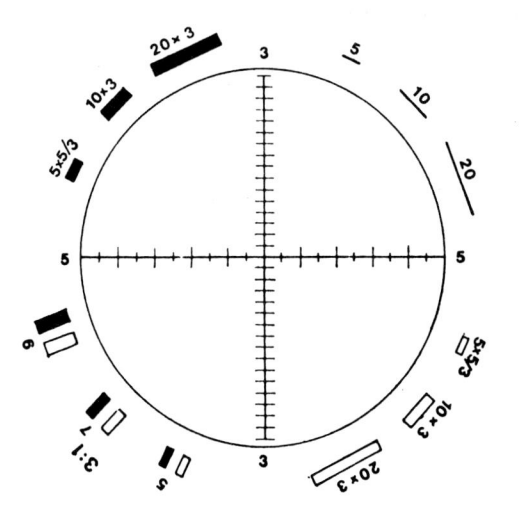

Fig. 1.18. Becket and Walton eyepiece graticule (H. Walton, *Annals of Occupational Hygiene*).

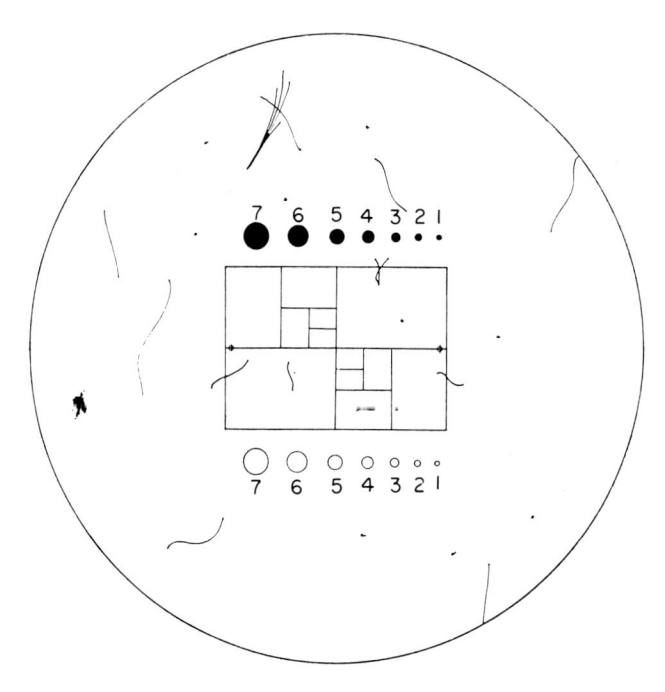

Fig. 1.19. BS 3625 eyepiece graticule (BS 3625 is reproduced by permission of the British Standards Institution 2 Park Street London W1A 2BS from whom complete copies of the standards can be obtained).

23

Chapter 1

identity and place it centrally over the blue circle marker.

3 With the micro-pipette drop 0.1ml of clearing fluid on the slide over the area marked by the blue circle.

4 Using blunt forceps carefully place the loaded membrane on to the area of clearing fluid, dust side uppermost; leave for three minutes.

5 Place a glass cover slip over the membrane and leave for at least 30 minutes.

To estimate the number of respirable fibres on the membrane, proceed as given below:

1 Set up phase contrast lighting conditions on the microscope and check the dimensions of the eyepiece graticule at 500× magnification by means of a stage micrometer. If the microscope is regularly used for counting then the size of the graticule will be known.

2 Place the slide on the stage of the microscope and focus at a low magnification to observe the distribution of fibres over the whole slide and to determine whether the fibres are evenly distributed or not.

3 Change to the high magnification (500× or thereabouts). If the fibres are evenly distributed then select fields randomly and count until 200 respirable fibres have been counted or if the slide contains few fibres, until 100 fields of view have been examined. Use the digital counters to record the number of fibres and the number of fields selected. Do not count the fields close to the outside edge of the filter as contamination can occur from the filter holder and give a false result.

Note. A respirable fibre is defined as: one that is greater than $5\mu m$ in length and having a length/breadth ratio of at least 3:1 and a diameter less than $3\mu m$. The blocks and lines around the outside of the Becket and Walton graticule in fig. 1.18 assist in selecting the correct fibres to count.

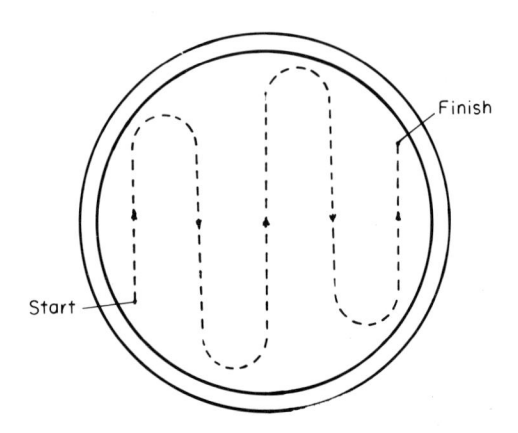

Fig. 1.20. Viewing pattern for scanning slide when counting asbestos.

Fibres which are overlapping the field of view should only be included if more than half their length is in the field.

4 If the sample is unevenly distributed or sparse in number then observe 100 fields selected in a viewing pattern as shown in the diagram in fig. 1.20.

Calculations

Let:
    effective diameter of the membrane (i.e. the actual diameter less the overlap due to the retaining ring of the filter holder) $= D$ mm
    the diameter of each field of view $= d$ mm
    the number of fields examined $= n$
    the number of fibres counted $= N$
    the volume of air sampled $= V$ ml
$V =$ flow rate of pump (1 min^{-1})× duration of sampling (min)×1000 ml

$$\text{Estimated number of fibres sampled} = \frac{D^2}{d^2} \times \frac{N}{n} \text{ fibres}$$

$$\text{Fibre concentration} = \frac{\text{estimated number of fibres}}{V} \text{ fibres ml}^{-1}$$

It is assumed that the type of asbestos has already been identified, possibly by an examination of a bulk sample from the source of contamination using a suitable technique such as polarized light or electron microscopy. Compare the results obtained with the standards laid down by the Health and Safety Executive as follows:

crocidilite fibres: 0.2 fibre ml^{-1} when measured over any 10 minute period.
other types: 2.0 fibres ml^{-1} when measured over a 4-hour period; short term exposure should not exceed 12 fibres/ml when measured over any 10 minute period.

The Health and Safety Commission set up a committee under the chairmanship of Mr W. Simpson to look into the risk to health from exposure to asbestos. This committee has made certain recommendations concerning new asbestos standards termed 'control limits' as follows:
crocidilite fibres: 0.2 fibres ml^{-1} for a four hour sampling period
chrysotile fibres: 1.0 fibre ml^{-1}
amosite fibres: 0.5 fibre ml^{-1}

Possible problems

1 The field may be heavily contaminated with other dust, thus making

Chapter 1 it difficult to see the fibres. In this case reduce the sampling flow rate or reduce the sampling time.

2 The ensure that the plane of view is correctly focussed, always use a membrane which has a gridded surface and focus on the grid lines.

3 Accuracy in counting is increased by spending as long as possible on each field.

4 Care must be taken to ensure absolute cleanliness throughout to prevent unwanted contamination of the sample.

5 Inter-person/laboratory counting should normally be employed to indicate individual variations in counting characteristics.

The choice of filter and filter holder to suit a specific dust, fume or mist

When sampling for a specific dust of known composition or type it is important to consult the analyst before starting as the method of analysis will vary for the different chemical composition of the dust. Each

Table 1.5. Details of filters and filter holders to be used for various types of dust

Type of dust	Method of analysis	Filter required*	Filter holder required
Asbestos fibres	Optical microscopy	Cellulose ester*	Open face
	Scanning electron microscopy	Nuclepore	Open face
	X-ray diffraction	Silver membrane	Open face
Man-made fibres	Optical microscopy	Cellulose ester*	Open face
	Gravimetric	Glass fibre	Modified UKAEA†
Silica	X-ray diffraction	Silver membrane	Cyclone
	Infra-red	Polyvinyl chloride	Cyclone
Lead, heavy metals, their oxides and salts	Atomic adsorption spectroscopy	Cellulose ester or Glass fibre	UKAEA with 4mm hole
Nuisance & general	Gravimetric	Glass fibre	Open face
	Optical microscopy	Cellulose ester	Open face
Unknown dusts	X-ray diffraction	Silver membrane	Open face
Coal	Gravimetric	Glass fibre	Cyclone or MRE 113
Coal with rock	Infra-red	Polyvinyl chloride	Cyclone or MRE 113
	X-ray diffraction	Silver membrane	Cyclone or MRE 113
Oil mists	Gravimetric	Glass fibre	Open face
	Fluorescent spectro-scopy	Cellulose ester*	Open face
Welding fume	Gravimetric	Cellulose ester*	Open face
	Atomic adsorption spectroscopy	Cellulose ester*	Open face

*Note that where cellulose ester membranes are used the pore size should be $0.8\mu m$.
†The modified UKAEA has a ring of six 4mm dia. holes on a 12.5mm pitched diameter around a central 4mm dia. hole shown in fig. 1.3.

method will require the dust to be presented in a particular way and the analyst will advise as to the best filter to use to suit the technique being applied. Given below is a table of some of the more commonly encountered dusts and the recommended filter to use.

Where size selection is being used such as a cyclone for respirable dust it is important to ensure that the correct air flow rate is passing to suit the type of selector being used and that the flow rate is not pulsating.

To take a sample, proceed as with the open face filter or the cyclone separator methods described previously but using the type of filter and holder detailed in table 1.5.

To trace the behaviour of a dust cloud using a Tyndall beam

Aim

Many particles of dust are too small to see with the naked eye under normal lighting conditions but when a beam of strong light is passed through a cloud of particles they reflect the light to the observer and as a result become readily visible. A natural occurrence of this phenomenon is observed when a shaft of sunlight shines into a dark building highlighting the airborne particles. The scientist Tyndall made use of this principle to observe the behaviour of dust clouds using a beam of light and his name has been associated with the technique from that time on. Thus if a portable lamp having a strong parallel beam is set up to shine through an environment suspected of being dusty the movement of the particles can be observed. Although no numerical measurements are normally made, the performance of extract ventilation systems associated with dust emitting processes can be watched and corrective designs made if unsatisfactory capture is noticed. It may be useful to film or video record the occurrence.

Equipment required

A strong parallel beamed lamp, mains or battery powered (a car foglamp is unsuitable but a spotlight is better), a tripod stand for mounting the lamp and a black screen. There is a Tyndall beam apparatus commercially available in a portable kit form which has its lamp reflector designed to provide a strong parallel beam of light.

Method

Set up the lamp and screen as shown in fig. 1.21 with the observer in the position indicated. When the lamp is switched on the dust cloud should be clearly seen and the movement of the particles observed, photographed or filmed. The best position of the lamp and screen may have to

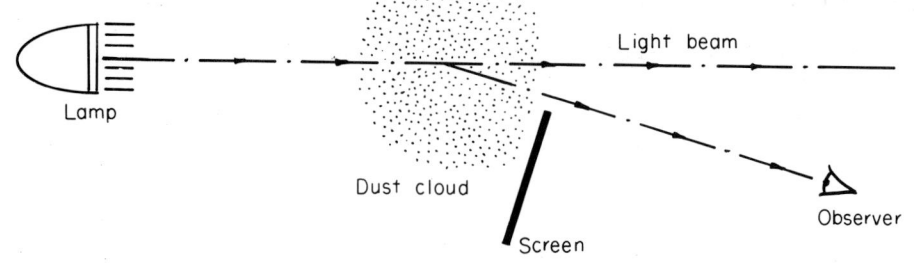

Fig. 1.21. Layout of Tyndall beam apparatus in relation to a dust cloud (A & G Marketing Ltd).

Fig. 1.22. Tyndall beam lamp kit (A & G Marketing Ltd).

be adjusted by trial and error but it is important to shield the lamp from the eyes of the observer or the camera to prevent glare.

Further reading

American Conference of Government Industrial Hygienists. *Air Sampling Instruments Manual.* ACGIH, Cincinatti, Ohio, 1977.

Lee GL. Sampling: principles, apparatus, surveys. In Waldron HA & Harrington JM (eds) *Occupational Hygiene.* Oxford: Blackwell Scientific Publications, 1980.

Health and Safety Executive, Guidance Note EH/10, *Asbestos Hygiene Standards and Measurements of Airborne Dust Concentrations.* HMSO, 1976.

Asbestos Research Council, Technical Note 1. *The Measurement of Airborne Asbestos Dust by the Membrane Filter Method.* Rochdale: ARC, 1971.

Health and Safety Executive. *Asbestos—Final report of the Advisory Committee* (Simpson). HMSO, 1980.

CHAPTER 2
GASES AND VAPOURS

Introduction

Many industrial processes as well as natural or biological degradation processes utilise or produce gas, often under pressure. Containing these gases is a major problem since many are toxic or may cause asphyxia in enclosed environments. Leaks may occur around joints, valves, or through piping, and access covers when opened release gas to the atmosphere. Therefore suitable precautions have to be taken to contain these gases and their presence has to be monitored.

Solvents are another source of respirable materials. These are liquids chosen for their ability to dissolve a particular material and often for their ability to evaporate and dissociate themselves from the solute. In so doing a vapour is formed and it is unfortunate that the majority of solvents have biological activities. Most have a narcotic effect acting upon the central nervous system slowing down nerve responses; others are extremely toxic or are sensitizing agents and some are even mutagenic and tumouregenic.

The range of gases and vapours that occur in industry, commerce, medicine, agriculture and in the home and street is vast, requiring a wide variety of monitoring and detection techniques. A few methods of sampling and detection are described in this chapter. The techniques available fall into two basic types: direct measurement with instruments or detection devices; and indirect analysis of air collected from the workplace and examined in the laboratory. The satisfactory use of many of the instruments available requires a good knowledge of chemistry and is best left to those trained in analytical chemistry, but colourimetric detector tubes and paper tape monitors can be used by any competent individual. Certain collection devices can also be straightforward to use provided the analysis is left to someone experienced.

As with airborne dust measurement, if a time-weighted average concentration is required then it is usual to collect a sample of the air over the working period rather than use a direct reading instrument which usually takes an instantaneous measurement or 'grab sample'. Some direct reading instruments, however, do have recorders attached to them so that peaks and troughs of concentration can be seen and averages calculated. Some are linked to computers so that printout facilities can provide information in whatever form the user has programmed. These devices produce general workroom levels of concentration as it is difficult to attach them to workers to monitor their personal breathing zone exposures. However there are now available specially designed long term detector tubes that can be used as personal samplers for certain pollutants over the working shift.

Sampling for airborne chemicals can be done in two ways: by collecting continuously a small amount of the workroom atmosphere gradually filling a container over the test period, or by allowing workroom air to pass through an adsorbent material such as activated charcoal which will adsorb certain gases which can later be desorbed in a more concentrated form for analysis. In the first case the receptacle will contain a sample of the workroom atmosphere which represents the average mixture of all gases present at the sampling position over the period sampled. Unless the sampling air flow rate is low the final sample can be quite bulky. With the adsorbent method of collection the pollutant is concentrated which makes analysis more reliable.

Equipment available

Collection devices

Before embarking upon any form of air sampling it is important to discuss with the analyst which method of collection best suits the analytical technique to be employed when the samples are returned to the laboratory. The method of analysis dictates the type of container or collection device to be used as the final sample has to be introduced into the analytical instrument.

Containers

If a sample of the workroom air is required for complete analysis, that is, for the normal constituents of fresh air, oxygen, nitrogen, and carbon dioxide; and for certain pollutants such as carbon monoxide, sulphur dioxide, oxides of nitrogen and some hydrocarbons such as methane, then it can be collected in a container to be returned to the laboratory. But if other gases and vapours are required to be known and they are sampled in a container there is a risk that some pollutant may be adsorbed on the walls of the container and not released when required for analysis and some may well be released at a later time when the container is being used again. Therefore there are some risks with this method and some knowledge of the behaviour of the gases to be sampled is required.

Samples can be collected in: syringes, glass pipettes, plastic or rubber bags, metal cylinders or evacuated vessels of glass or metal. For grab samples they can be filled by hand pump or in the case of evacuated vessels by releasing the vacuum to allow air to enter. For long term samples bags should be used being filled slowly over the sampling period by a small battery powered pump. There is an evacuated cylinder

Gases and Vapours

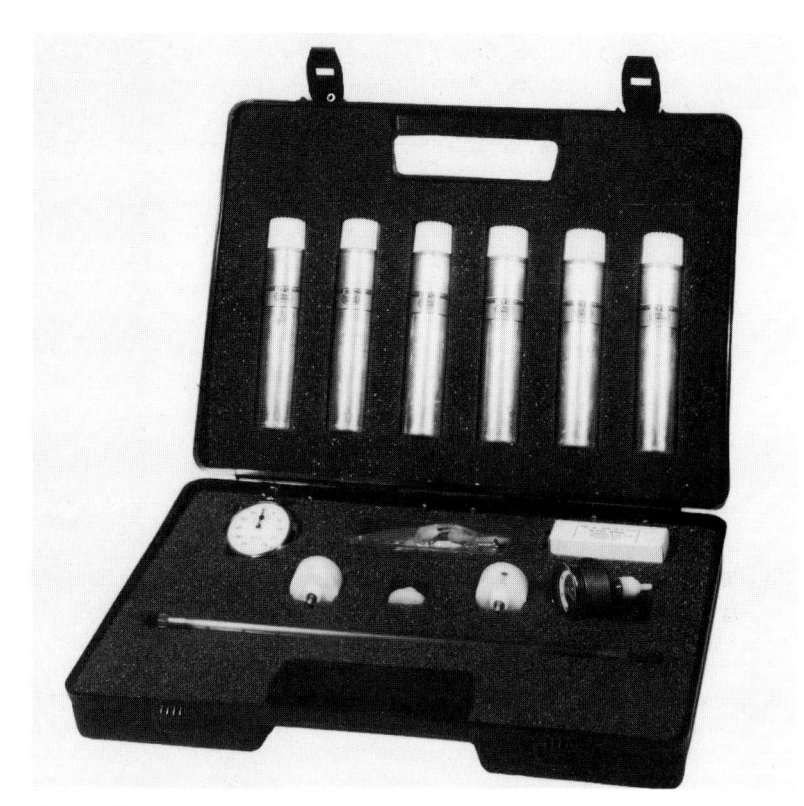

Fig. 2.1. Vacuum operated personal sampler kit (Casella Ltd).

marketed by Casella Ltd, illustrated in fig. 2.1, which is fitted with a release valve that allows air to enter slowly over an eight hour period.

One disadvantage with this method of sampling is that with the exception of the vacuum tube gas sampler it is difficult to obtain a personal sample as most receptacles are rather bulky or are only suited to grab sampling. Another disadvantage is that the sample collected contains the pollutant at the average concentration at which it occurs at the sampling point. This may be very low thus the analytical equipment used to determine the concentration must be very sensitive.

Adsorption methods

Certain gases and vapours are readily adsorbed by solid materials such as silica gel, activated charcoal and various types of porous resin. When air containing those gases or vapours is passed through the material they will be adsorbed. Tubes of metal or glass can be made or purchased containing one of these materials although charcoal is the most common. The

Chapter 2

tubes should remained sealed until ready for use.* When a continuous stream of air is passed through them by means of an air pump the material will adsorb those gases or vapours for which they are designed. Provided that the amount of air passed through has not overloaded the material then the amount of pollutant collected can be determined back in the laboratory and an average concentration calculated knowing the amount of air passed through.

One advantage of this technique is that the pumps and tubes are small and can be attached to a worker in a similar way to the airborne dust samplers. Another advantage is that the pollutant is collected in a concentrated form thus analytical techniques need be less sensitive than with the container methods and several gases or vapours can be collected and analysed from the same tube.

Not all gases or vapours can be collected by means of the adsorbent tube. Another technique available is to bubble the workroom air through water or some liquid in which the gas is soluble. The analyst then handles the gas in solution. The container for the liquid is known as a 'bubbler' because the sampled air is introduced via a tube whose end is below the surface of the liquid thus the air is bubbled through it. Unfortunately most bubblers require to be kept upright to prevent the reagent from spilling out. This limits their use for personal sampling because workers wearing them would not be able to bend or stoop. Another disadvantage of this technique is that it is difficult to guarantee that all the gas passing through the bubbler will be absorbed as bubbling does not ensure 100% contact between the gas and the liquid. To improve contact 'fritted bubblers' are used which have an inlet tube whose end is of porous glass producing very small bubbles.

Passive samplers

Recently techniques have been developed whereby adsorbent material can be used to sample concentrations of airborne pollutants without using a pump to draw air through the collector. The adsorbent material is contained in a holder designed to allow the gases to diffuse and/or permeate to the adsorbent surface. These holders are small enough to be worn like a lapel badge and are free of any pump or tubing, see fig. 2.2. At the end of the sampling period the holder is returned to the laboratory where the adsorbent material is removed and the amount of gas or vapour collected can be analysed as with adsorbent tubes.

Much research continues to be undertaken to try to relate the amount of sample collected with the true airborne concentration as it is difficult to establish how much of the workplace air has come into contact with the adsorbent material. Current opinion believes that passive samplers do not provide a very precise measure of the workplace

---

*Adsorbent tubes must not be confused with colourimetric detector tubes.

Gases and Vapours

Fig. 2.2. Passive sampler (D.A. Pitman Ltd).

concentration but can be used to indicate the order of magnitude of the levels particularly if used routinely in conjunction with some pump and tube samplers.

Pumps

When slowly filling bags and for drawing air through adsorbent tubes a low air flow rate is required. Pumps are available which provide rates as low as 2 millilitres per minute (ml min^{-1}) but the range for this type of sampling is between 2 and 500ml min^{-1}. These pumps operate under the same principles as the dust sampling pumps, that is, diaphragm, piston and rotary, and, with certain exceptions, are supplied by the same manufacturers. The correct flow rate to use is influenced by the airborne

Chapter 2

concentration of the pollutant, the ability of the material to adsorb, the size of the sampling tube and the duration of sampling. Bubblers tend to require higher flow rates therefore the medium flow pumps as used in dust sampling are to be prefered.

Most pumps available have a variable flow control either by means of varying the stroke of the diaphragm or by varying the speed of the driving motor. One pump, however, operates by providing a suction through a 'critical orifice' which limits the amount of air flowing through it, a change in flow rate being achieved by changing the size of the orifice. Flow rates of any pump should always be checked against a soap bubble flow meter or a calibrated rotameter.

Tube holders

In order to hold the adsorbent tube and to attach it to a worker if required, a tube holder is available from the suppliers of the low flow rate pumps. A piece of flexible plastic tubing connects the holder to the pump. An example of a tube holder containing a charcoal adsorbent tube is shown in fig. 2.3.

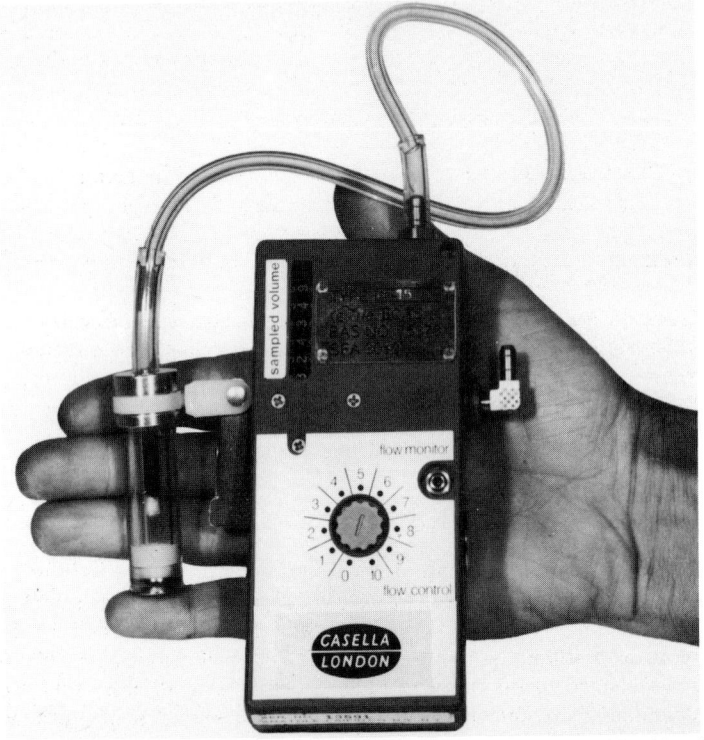

Fig. 2.3. Low flow rate sampling pump and adsorbent tube in holder (Casella Ltd).

Adsorbent tubes

The materials which are used for adsorbing gases and vapours are listed as follows: charcoal, silica gel, alumina, various porous polymers under the names of; Poropak P, Poropak Q, Chromosorb 101, Chromosorb 102 and Tenax GC. They each have the ability to adsorb various gases and vapours but charcoal is the most widely used as it has an adsorbent affinity with more substances than any other material. The materials are contained in either glass or metal tubes depending upon the method of desorbing the gas in the laboratory. One method involves breaking the tube so that the adsorbent material falls into a liquid which leaches the gas out of the material into solution. Glass tubes are used in this application. Another method involves heating the tube to drive off the collected gas into a detection system. Metal tubes are used for this technique. Tubes can be made up in the laboratory or purchased from the suppliers of sampling pumps.

The tubes designed for solvent leaching are made up in two sections to indicate what is known as 'breakthrough'. This occurs when the sorbent material is completely saturated with the sampled gases. If the second section of the tube is free of the sampled gases then breakthrough has not occurred in the first section and the total sample collected on that section can be used in the calculation for airborne concentration. Breakthrough occurs when the sample volume is too high or the sample duration is too long to suit the airborne concentration in the sampling position.

With metal tubes for thermal desorption no second section can be applied thus breakthrough must be carefully guarded against by calculation of breakthrough volumes not to be exceeded during sampling. For more details the book by W. Thain mentioned in further reading should be consulted.

It cannot be emphasised too strongly that when sampling for airborne gases and vapours the analyst who undertakes the analysis of the collected samples must advise as to the best method and, if adsorbent tubes are to be used, what sorbent material and what sample volumes to use.

Colourimetric detector tubes

A comparatively simple method of detection of airborne gases is by using the detector tube. This consists of a glass tube containing crystals treated with a chemical reagent which will react with a particular gas and change colour as a result. As contaminated air is drawn through the tube a colour change occurs from the inlet end which extends along the tube depending upon the concentration of the gas present. A scale is printed on the side of the tube on which the measured concentration is indicated against the length of stain for a particular sample volume. A hand

Chapter 2

operated suction pump is provided to ensure the correct sample volume if operated according to the makers instructions. The tubes are sealed at each end and both ends must be broken before inserting into the pump. An example of the type of sampler is shown in fig. 2.4. Two companies specialise in these devices: Dräger and Gastec and between them detector tubes for over 200 substances are available. The tubes supplied by these companies are not interchangeable, that is, Gastec tubes cannot be used with a Dräger pump and vice-versa.

Whilst the technique appears simple there are certain difficulties which can lead to error in the results obtained. The detector tubes deteriorate with time and have a shelf life of no more than two years if stored at normal room temperature. The presence of other gases can interfere with the gas to be measured. The manufacturer will advise in these cases. Also it is important that the designed sample volume passes through the tube otherwise the result will be invalid. This means that any damage to the pump causing a leak in the airflow will reduce the volume of air passing through the tube.

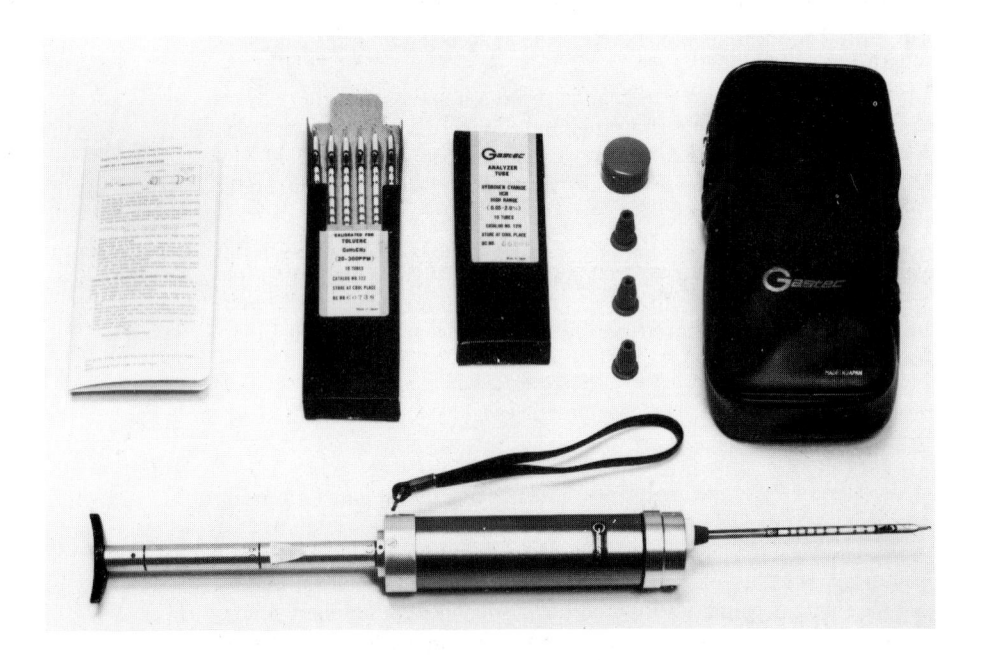

Fig. 2.4. Colourimetric detector tube kit (D.A. Pitman Ltd).

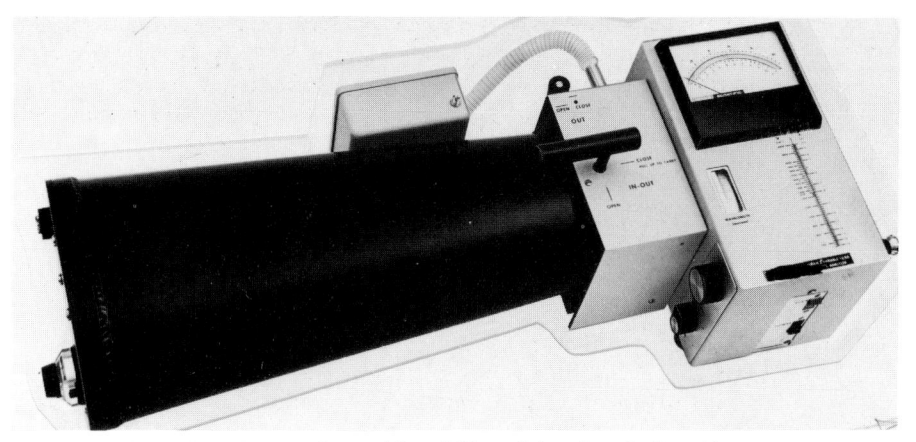

Fig. 2.5. Miran infra-red gas analyser with variable path length and adjustable wavelength.

Direct reading instruments

Many gases and vapours can be detected by direct reading instruments. Some are limited to a specific gas, such as, carbon monoxide, sulphur dioxide, mercury vapour and many others. Some can be tuned to measure many different gases using an infra-red beam of variable wavelength (fig. 2.5). Others can detect the concentration of a mixture of gases or a range within a group or family of compounds but is not specific to any one. This instrument has the ability to be specific by the addition of a gas chromatographic column which then makes it a small portable gas chromatograph. There is also a type of instrument illustrated in fig. 2.6 which draws air through a paper tape which is impregnated with a chemical reagent which reacts with certain specific gases to produce a stain whose intensity depends upon the airborne concentration. By shining a beam of light through the stain on to a light sensitive cell the intensity of the stain can be measured and the concentration of the gas estimated. This instrument can be used to detect several gases by changing the type of impregnated paper tape to suite the specific gas. The colour change principle is also used in an instrument which uses liquid reagents rather than paper tape.

General

The range of instruments available is so large that it is impossible to cover them in a book of this nature. The instruments specific to particular gases and the paper tape types are generally easy to use and good instructions are always provided by the manufacturers. The tunable infra-red instruments and the portable chromatograph types require a good knowledge of chemistry to obtain reliable results.

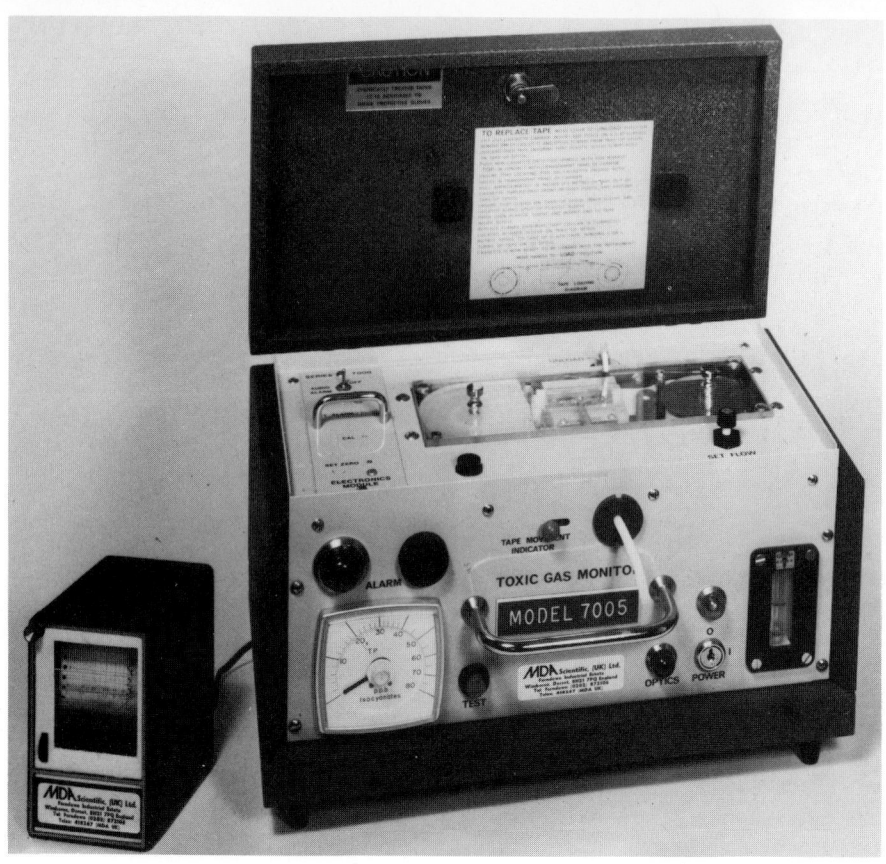

Fig. 2.6. Paper tape sampler (MDA Scientific (UK) Ltd).

To obtain a personal sample for solvent vapours using an adsorbent tube

Purpose

In order to obtain a time weighted average concentration of the exposure of a worker to solvent vapours it is necessary to collect a representative sample of the air from his breathing zone over the period of exposure. This can be achieved by adsorbing the vapour on to a medium such as charcoal. Other media are available but the analytical chemist will advise as to the most suitable for the vapours in question.

Equipment required

A low flow rate pump, an adsorbent tube, a tube holder with a length of plastic tubing to connect to the pump, a calibrated rotameter, a harness may also be useful.

Method

1 Break the glass seals or remove the covers at each end of the tube packed with adsorbent and insert into the holder. Connect the other end of the tube to the pump.
2 Attach the tube holder to the worker as close to the breathing zone as possible, the lapel or clothing close to the collar bone being the most acceptable place. Place the pump in a convenient pocket or hang from the worker's belt.
3 Turn on the pump and note the time of starting.
4 Some pumps are fitted with a stroke counter and the reading on the counter must be noted before starting. Also these are fitted with an orifice to provide the correct flow rate. Check that this orifice is the one recommended by the manufacturer for the flow rate required. With other types of pump check the flow rate with a calibrated flow meter and note its value. The analyst should have advised on what flow rate to use knowing the concentration of vapours expected and the length of the sampling period.
5 From time to time during the operation check the flow rate using the rotameter.
6 At the end of the period stop the pump and note the time and the reading on the stroke counter where applicable.
7 Remove the apparatus from the worker and place seals at the open ends of the adsorbent tube and label the tube for identification purposes.

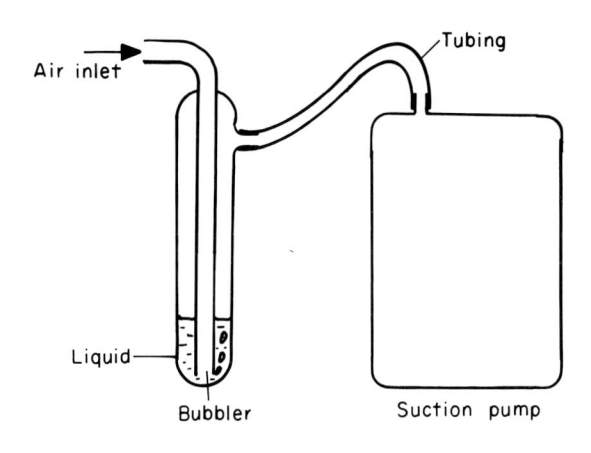

Fig. 2.7. Bubbler sampling train.

8 Send the tube to the analyst and remind him of the substances likely to be found adsorbed on the medium and tell him for which substances he is required to analyse.

Calculations

It is first necessary to establish how much air has passed through the tube. The calculation to do this will depend on the type of pump used. With pumps having a stroke counter ascertain the total number of strokes that occurred during the sampling period by subtracting the first reading from the second and multiply the result by the displacement value of each stroke as advised by the manufacturer for the setting or orifice used. Convert the results to cubic metres, that is, if the result is in ml min^{-1} multiply that value by the total number of minutes that elapsed during the sampling period and divide by 1 000 000.*

After the adsorbent tube has been analysed the total amount of each vapour tested will be given in milligrams (mg) or micrograms (μg). Divide the weight given by the total sampled volume to give the concentration in mg m^{-3} or μg m^{-3}.

Example

A pump passing 5ml min^{-1} sampled for 6 hours 40 minutes using a charcoal adsorbing tube in a workplace polluted with paint solvents. The analyst reported that the sample contained the following amounts of solvent: 1-butanol, 0.156mg; xylene, 0.298mg; stryrene, 0.187mg. Calculate the airborne concentrations of each.

Total time of sampling = 6×60+40 = 400 min
Total airflow through tube at 5ml min^{-1} = 5×400 = 2000ml = 0.002m^3*
Concentrations:

$$1\text{-butanol} = \frac{0.156}{0.002} = 78\text{mg m}^{-3},$$

$$\text{Xylene} = \frac{0.298}{0.002} = 149\text{mg m}^{-3},$$

$$\text{Styrene} = \frac{0.187}{0.002} = 93.5\text{mg m}^{-3}.$$

Possible problems

The problem of 'breakthrough' must be guarded against. This occurs when the adsorbent material has been overloaded. After sampling for a

---

*It must be pointed out that the most inaccurate part of this or any similar test lies with the flow rate of the pump which could be as much as ±10% in error. Therefore it is unwise to be pedantic about the last figure in the result.

period of time the adsorbent can become saturated with the vapour and no more can be collected thus making continued sampling pointless. It is not always possible to know when this happens. It may occur when the sampling rate is too high or the concentration of vapours upon which the rate has been based has been under-estimated. Some proprietary tubes contain a second stage which is analysed separately from the first. If this second stage contains no adsorbed vapour then breakthrough has not occurred and the first stage holds all the sample. If there is some doubt about the range of concentration that is likely to occur in the workplace, therefore giving rise to uncertainty as to whether breakthrough could occur, then a second adsorbent tube should be added to the first in series by means of a short piece of plastic tubing. Each tube should then be analysed separately. A further precaution can be employed by adding two 'two stage' tubes in series thus providing four stages of adsorption. Provided that the last stage is free of vapour when analysed then it can be confidently assumed that all the vapour has been collected in the previous stages. The total vapour collection on all stages must be used in the calculation. It is important to consult the analyst about breakthrough as it is possible to estimate sampling times to prevent this occuring provided the vapours to be collected are known in composition and a reasonable estimate of the airborne concentration known.

The collection of gases using a sampling bag

Purpose

It is possible to obtain a time weighted average concentration of an airborne gas using a direct reading instrument if a bag is filled slowly over a timed period. The resulting mixture in the bag will be a mean of the peaks and troughs of the concentrations occuring during the period. Therefore if some of the collected sample is introduced into the direct reading instrument the time weighted average concentration will be indicated. Due to the bulk of the bag it is difficult, but not impossible, to obtain a personal sample.

Equipment required

A low flow rate sampling pump, a bag and a length of connecting tubing. As most sampling pumps are designed for suction only a few are fitted with a discharge nozzle to which tubing can be attached. For this operation it is essential to choose a pump with that facility. With certain makes of pump it is possible to fit a nozzle to the discharge port by soldering or using a strong adhesive.

The flow rate should be as low as possible in order to make the sampling period as long as possible and to keep the size of the bag to manageable proportions. If the pump is capable of a flow rate of, say,

Chapter 2

5ml min^{-1} then during the course of eight hours it will have passed some 2400ml of sample or 2.4 litre which represents a bag of approximately 13.5cm cube. However if the pump handles 500ml min^{-1} for eight hours the sample would amount to 240 litre requiring a bag of 62cm cube which would be much more difficult to manage. Table 2.1 gives the total volume collected for various periods and for various flow rates.

Table 2.1. Total collected volume of sample for various flow rates and sampling duration.

Flow rate ml min^{-1}	1 hour	2 hour	4 hour	8 hour	Remarks on manageability
2	0.12	0.24	0.48	0.96	Could be worn by a worker
5	0.30	0.60	1.20	2.40	
10	0.60	1.20	2.40	4.80	Too large to be worn by a worker but one person could transport four
50	3.00	6.00	12.00	24.00	
100	6.00	12.00	24.00	48.00	Much too large to be worn and only one could be reasonably transported by one person
200	12.00	24.00	48.00	96.00	
500	30.00	60.00	120.00	240.00	
1000	60.00	120.00	240.00	480.00	

The bags should be of strong non-porous plastic with welded seams and a single supply tube with some means of closing it. Specially designed sampling bags are available made of a variety of plastics some having aluminium impregnated into the pores to minimise leakage and surface retention.

Method

1 Completely deflate the bag having ensured that it has been flushed through at least four times with clean air to remove or dilute any previous sample.
2 If no valve is fitted to the supply tube then fit a tube clamp to it.
3 Connect the bag to the discharge nozzle of the pump via suitably sized tubing.
4 Set up the assembly at the point of sampling, either on a worker if a very low flow rate pump is being used or in a static position if not. For static sampling it is useful to attach the pump and tubing to a tripod by

means of adhesive tape. For personal sampling a bag holder and harness are available.

5 Open the bag valve or unclamp the supply tube and start the pump noting the time.

6 Adjust the flow rate of the pump to suit the size of the bag and the sampling period although it is not important to know the exact flow rate.

7 Stop the pump after the sampling period is completed and seal the valve or clamp the tube, noting the time.

8 Immediately transport the bag to the direct reading instrument which should have been warmed up and zeroed and calibrated in readiness. The instrument preparations should be done strictly in accordance with the manufacturers instructions. It is important to analyse the sample as quickly as possible as even the best sampling bags are slightly porous and the sample can diffuse away or change its composition.

Results

The reading obtained on the instrument is the time weighted average concentration of the pollutant measured.

Sampling for gases using a bubbler

Purpose

Some gases are not readily adsorbed by solids but dissolve in liquids or form chemical reactions with certain reagents when in contact. One method of introduction is to bubble the workroom air through the appropriate liquid using a device called a bubbler. Time weighted average concentrations can be obtained by allowing the bubbling action to continue for a period of time. The success of this technique depends upon the readiness of the pollutant in question to react with the solution or reagent. Unfortunately the nature of bubbles is such that not all the gas may come into contact with the liquid thus some of the pollutant can escape. To overcome this some bubblers are designed to produce very fine bubbles using a device known as a frit. Also bubblers can be staged in series so that as the air leaves the first it then passes into a second or even a third.

Clearly devices such as these are unwieldy to use particularly if a personal sample is to be obtained from the breathing zone of a worker and there is a risk that the liquid may be spilled if not kept upright. A spill-proof bubbler has recently been designed. Nevertheless there is obvious resistance by workpeople to wear devices containing what may be dangerous chemicals, therefore bubbling techniques are usually confined to static workplace sampling rather than personal.

Chapter 2   *Equipment required*

A bubbler containing the correct amount of appropriate liquid as advised by the analytical chemist, a suction pump, some connecting tubing, a tripod stand, some adhesive tape, a calibrated rotameter, and, if the atmosphere to be sampled is dusty, then an open face filter holder containing a glass fibre filter should be added at the entrance to the sampling train.

Method

In order to establish the required pump flow rate it is important to have a trial run beforehand using clean water in the bubbler the amount being the same as that required for the test. This is to ensure that no liquid is unintentionally drawn into the pump thus causing damage. If the bubbler is filled beforehand and carried to the site it must be kept upright and the open ends of its tubes sealed. It may be more convenient to fill on site but that involves carrying a measuring device to meter the correct amount of liquid.

1   Connect the bubbler to the pump as shown in fig. 2.7, that is, connect the pump to the tube not in contact with the liquid. *Do not connect the pump to the central tube.*
2   At the sampling site attach the assembly to the tripod stand by means of adhesive tape or other suitable means.
3   Start the pump at the preset rate of flow noting the time.
4   Using a calibrated flow meter check and note the air flow rate passing through the train and repeat from time to time throughout the test.
5   At the end of the sampling period stop the pump and note the time.
6   Disconnect the tubing from the bubbler and seal its ends and label it.
7   Return the bubbler to the laboratory immediately so that analysis can proceed without delay as some chemicals can change in a short period of time.

Calculation

Establish the total amount of air that has passed through the sampler by multiplying the elapsed time by the flow. Convert to cubic metres (m³). The analyst will report the amount of pollutant collected in milligrammes (mg) or microgrammes (μg) and from that the airborne concentration can be obtained in mg m^{-3} or μg m^{-3} by dividing the sample amount by the total flow.

Example

A bubbler was run for 20 minutes at a flow rate of 1.5 litre min^{-1} in an atmosphere containing formaldehyde. The analyst reported that the sample contained 0.048mg of formaldehyde. Determine the airborne concentration.

Total airflow through the bubbler = $20 \times 1.5 = 30$ litre = $0.03 m^3$

Airborne concentration of formaldehyde = $\dfrac{0.048}{0.03}$ = 1.6mg m^{-3}

To measure the airborne concentration of a gas using a colourimetric detector tube

Aim

It is often necessary to obtain a quick indication of an airborne concentration of a gas. This may be required in a situation such as checking the concentration in an enclosed space, a sump or a large empty vessel before permitting persons to enter. It is also useful in a workplace or workroom to check the general concentration of a specific gas from time to time. Short term detector tubes can be used for this purpose but it must be remembered that these devices give an indication over a short period of time, usually less than one minute. The exact time taken to produce the result depends upon the type of gas to be detected and type of tube being used and is governed by the flow resistance of the tube. In occupational hygiene parlance this is known as a 'grab' sample and does not provide a time weighted average concentration. For certain gases a long term detector tube is available which is used in a long term sampling apparatus and which should not be confused with a short term tube.

Equipment required

A colourimetric sampling kit containing a hand operated suction pump and, from the same manufacturer, a box of detector tubes for the gas to be measured and of the range of concentration likely to be found. If sumps and enclosed vessels or if difficult access places are to be sampled then an extension hose is also required.

There are two basic types of suction pump available, the bellows type as shown in fig. 2.8 or a piston type as shown in fig. 2.3. Each pump is designed to pass a measured volume of air for an operating stroke, the bellows type passing 100ml and the piston type having two possibilities: namely a full stroke at 100ml and a half stroke at 50ml. Thus it is important to ensure that the correct tube is used with the pump for which it is designed.

The type of detector tube available depends upon the gas to be measured which dictates the chemical reaction which occurs within the tube in order to produce the indicating stain. With some types the concentration is indicated by means of the length of a coloured stain

Chapter 2

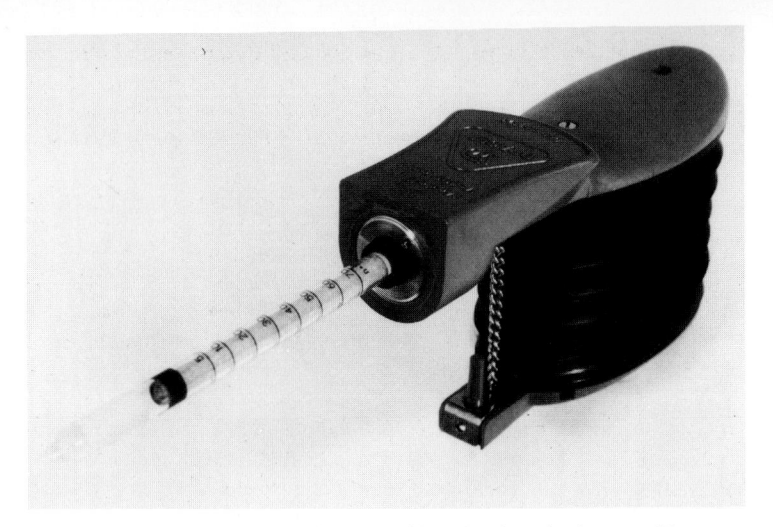

Fig. 2.8. Bellows type pump for use with colourimetric detector (Draeger Safety).

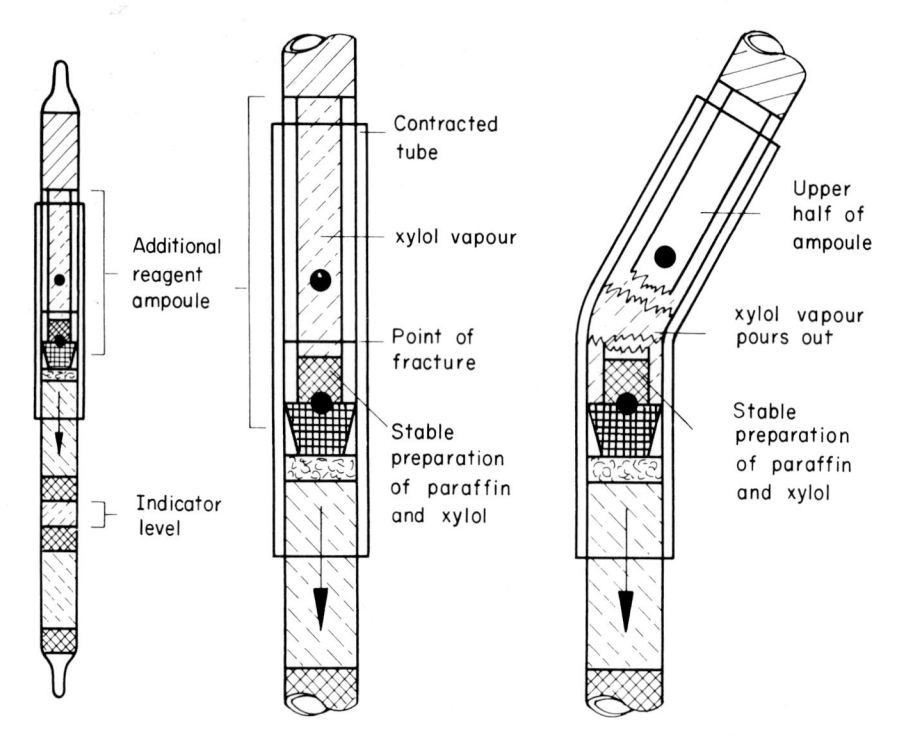

Fig. 2.9. Detector tube with ampoule (Draeger Safety).

measured against a single scale inscribed on the glass wall of the tube whereas a double scale is provided on some to accommodate two different numbers of pump stroke. Indication of concentration may also be done by colour change rather than length of stain a colour comparison being provided either built into the tube or as a separate item.

The construction of the tubes also vary for different gases depending upon the chemical reaction producing the stain. For example it may be necessary to activate the tube by breaking an ampoule of reagent within the tube as shown in fig. 2.9, this ampoule containing either dry powder, a liquid or a vapour. Alternatively it may be necessary to employ a pre-tube before the indicator thus placing the two tubes in tandem with a short connection between them.

Method

Using the bellows pump

Before making a test it is necessary to undertake some preliminary checks on the pump.

1   To check for a leak in the bellows: without breaking the ends of the glass detector tube insert it into the pump orifice and squeeze the bellows closed and release immediately. If the bellows remain closed then no leaks exist, a leaking pump would open during the test.

2   To check for a blockage in the suction channels: squeeze the bellows closed and with no tube in position the bellows should spring open immediately on release. If the channels are blocked the bellows would open relatively slowly.

To undertake the test read carefully the instructions supplied with the box of detector tubes.

3   Break both glass end seals on the detector tube using the tip breaking device on the pump. If an internal ampoule is provided, break that as per instructions.

4   Insert the tube into the suction orifice of the pump making sure the arrow on the tube points towards the pump. If a pre-tube is required, break its seals and connect it to the assembly according to the instructions.

5   Squeeze the bellows and release immediately, they will open at a rate governed by the flow resistance of the tube. Do not hinder this operation by trying to control the rate of opening. The bellows are fully open when the limit chain is taut. If sufficient gas is present in the air sampled a dark stain will appear from the zero and extending up the tube in response to the concentration of the gas present, see fig. 2.10.

6   If the scale on the tube requires one stroke then read the indicated concentration corresponding to the end of the stain as inscribed on the wall of the tube.

7   If the scale requires more pumps repeat operation 5, carefully count-

Chapter 2

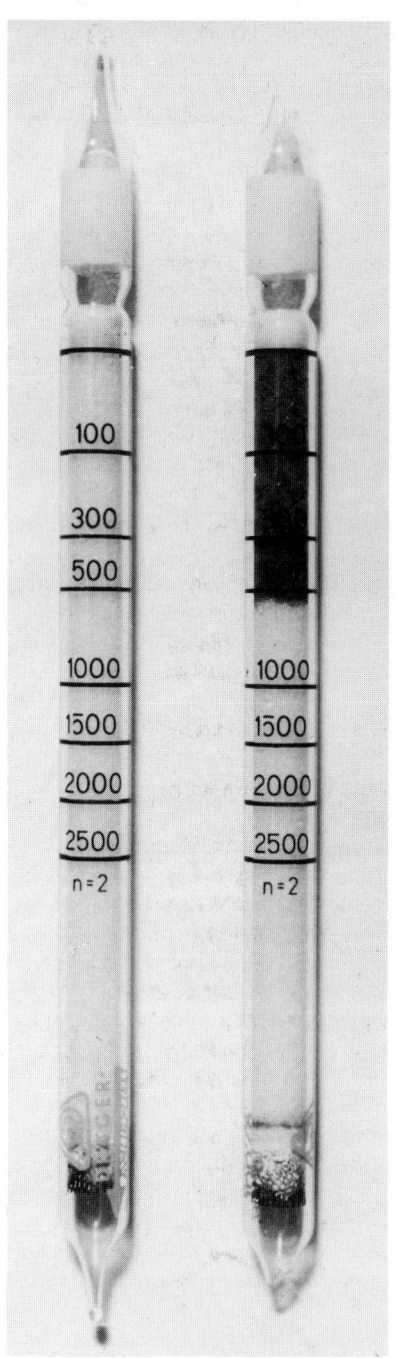

Fig. 2.10. Used and un-used detector tube showing staining (Draeger Safety).

ing the number of strokes until the required number is reached. The range of some tubes can be extended by increasing the number of strokes but this must be done according to the makers instructions. A stroke counter is available as an optional extra. At the end of the test read the concentration as described above.

Gases and Vapours

Using the piston pump

Before making the test it is necessary to undertake some preliminary checks on the pump.

1 To check the valves for leakage: without breaking the ends of the glass detector tube insert it into the pump orifice, move the pump handle so that the two red dots are not in line and pull out the handle several times fully quite quickly, then pull out the handle just 6 mm (¼ inch), hold for two minutes and release. If the handle returns to within 1.5mm (¹/₁₆ inch) of the closed position the valves are in order. If not the valves require to be lubricated according to the manufactures instructions.

2 To check the field volume: insert an unbroken tube as before, align the red dots and pull out the handle fully until it locks in the open position, wait for one minute then twist the handle a quarter turn to release it. If it is in order it should return to within 6mm (¼ inch) or less of the fully closed position but do not allow it to spring back but guide it gently as it returns. If it does not return as described lubrication is required as instructed by the manufacturer.

To undertake the test read carefully the instructions supplied with the box of detector tubes.

3 Break both glass seals on the detector tube using the tip breaking orifice provided on the pump.

4 Insert the tube into the pump making sure that the arrow on the tube is pointing towards the pump. If twin tubes are used connect the ends marked with a C by means of a short length of rubber tubing.

5 Push the pump handle fully in, align the red dots and pull out to the desired stroke position either half way or fully out according to the requirements of the test as instructed by the makers. The handle will lock in position and not return. If sufficient gas is present in the air sampled a dark stain will appear from the zero end extending up the tube in response to the gas concentration present. Do not release the handle until the stain has stopped extending.

6 The concentration can be read as that inscribed on the wall of the detector tube corresponding to the end of the stain.

7 If the tube requires more strokes of the pump as indicated by the makers in the instructions sheet, then twist the handle one quarter of a turn to release the locked position and repeat operation 5 for the required number of strokes. Carefully note the number of strokes. Read the concentration as instructed.

49

Chapter 2

Fig. 2.11. Long term detector tube in holder (Casella Ltd).

General

The manufacturers of detector tubes publish comprehensive handbooks outlining the main features of all their tubes and give guidance on where inaccuracies could occur. There are some important general remarks which should be made with regard to detector tube sampling in order to guide the user and to ensure that the results obtained are seen in perspective.

1 The results obtained for a specific gas can be considered reliable provided that no other gas is present which could interfere with the chemical reaction taking place inside the tube to produce the stain. The handbooks refer to interference gases. Therefore it is important that no test is undertaken without consulting the handbook.

2 The pump must be in good condition with a good seal between the tube and the pump suction orifice.

3 The tubes must be not more than two years old and preferably should be stored in a domestic type refrigerator. Dates are given on the box.

4 Detector tubes are designed to operate at 20°C and normal atmospheric pressure (1013mb, 760mmHg) and 50% relative humidity. In most cases a wide range of variation from these conditions can be tolerated but where tubes are sensitive to these factors correction charts are provided in the box supplied.

5 It is important to use a fresh tube for each test. If no stain appears after a test that tube must be discarded and not re-used.

Further reading

Thain W, *Monitoring Toxic Gases in the Atmosphere for Hygiene and Pollution Control*. Oxford: Pergamon, 1980.

American Conference of Government Industrial Hygienists, *Air Sampling Instruments Manual*, p. R1–R17, U1–U164, Cincinnati, 1978.

American Conference of Government Industrial Hygienists, *Analytical Methods Manual*, Cincinnati, 1980.

Lee GL, Sampling: principles, apparatus, surveys. In Waldron HA & Harrington JM (eds), *Occupational Hygiene*. Oxford: Blackwell Scientific Publications, 1980.

Health and Safety Executive, *Guidance Note EH15*. HMSO, 1980.

CHAPTER 3
HEAT

Introduction

The human body generates heat as a result of the burning of fuel. If that heat is dissipated too slowly deep body temperatures will rise, the opposite occurring if the outward heat flow is too fast. The rate of heat transfer between the body and its surroundings depends upon the thermal environment in contact with the skin. In order to effect a heat exchange between the two environments, that is, inside and outside the body, the normal heat transfer mechanisms of: conduction, convection, evaporation and radiation take place. Of these, convection and evaporation are interrelated and play a major role in dissipating body heat, therefore the temperature and the moisture content of the air are important parameters to measure. This is done by taking the wet and dry bulb temperatures of the air in the workplace. The study of the relationship between air temperature and moisture content is called 'psychrometry', that is the study of the behaviour of dry air and water vapour mixtures. Some further details of this topic are given later.

The radiant heat exchange between a person and the surroundings also plays an important part in the regulation of body heat flow as the skin radiates heat to colder surfaces and receives radiant heat from hotter surfaces. The rate of radiant heat flow is proportional to the difference in the fourth power of the absolute temperatures of the surfaces exchanging heat. The exact equation for the human body and its surroundings is difficult to establish because there are many surfaces at different temperatures, of different emissivities and subtending different solid angles to the body surfaces which, in turn, are constantly changing as the person moves. To integrate all these, the concept of 'mean radiant temperature' is used. This is defined as: the hypothetical temperature of a uniform black enclosure which exchanges the same amount of heat with the body as the non-uniform enclosure. The globe thermometer can provide a good indication of the radiant heat exchange likely to be found at a point although it is affected by air velocity and therefore does not provide the true mean radiant temperature.

Rates of heat convection and evaporation are also affected by the movement of air around the body, therefore it is important to measure the air speed at a workplace. Thus the four parameters which must be tested to obtain a true indication of the thermal environment are: dry bulb, wet bulb and globe temperatures, and the air velocity.

When any one or more of these parameters is excessive the heat flow to or from the body will be out of balance resulting in an uncomfortable or stressful situation occurring. Many indoor workplaces display unsatisfactory thermal environmental conditions for example: high radiant sources can be found in steelworks and glass making; high humidities in laundries, kitchens and deep wet mines; and cold conditions in deep-freeze stores and warehouses. Extremes of heat and cold are experienced at some time of the year in many outdoor work stations with regard to radiant heat, hot and cold air temperature, high and low humidities and high winds.

Equipment available

Dry bulb thermometers

These are normally mercury in glass thermometers with a variety of ranges to suit the environment to be measured. A useful range for indoor work is 5° to 65°C and for most outdoor work stations in Great Britain, −15° to 40°C. The small thermometers (scale length 100mm) measure to about 0.5°C having scale divisions of that order but the larger instruments (scale length 175mm) can be read to increments of 0.1°C although they are usually marked with the same divisions as the smaller ones.

There are also a variety of electrical thermometers available using either; thermocouple, diode or platinum resistance principles.

Wet bulb thermometers

These are simply dry bulb instruments as described above but with the bulb covered in a clean cotton wick wetted with distilled water. As the water evaporates from the wick heat will be removed from the bulb thus reducing the indicated temperature to below that of the dry bulb unless the air is fully saturated with water vapour and none can evaporate. The bulb can either be ventilated by an induced air current or it can rely upon natural air currents to remove the evaporated water vapour. In the latter case the reading on the scale is referred to as the 'natural wet bulb temperature'.

It is also possible to fit a wetted wick to certain electrical temperature indicators.

Sling psychrometer (whirling hygrometer)

This consists of a wet and dry bulb thermometer mounted in a frame with a swivel at the top end so that it can be rotated by hand to induce an air current to flow over the bulbs. The latest instruments have a distilled water reservoir into which the wick covering the wet bulb dips to provide a continuous flow of water (see fig. 3.1). Relative humidity can be

Chapter 3

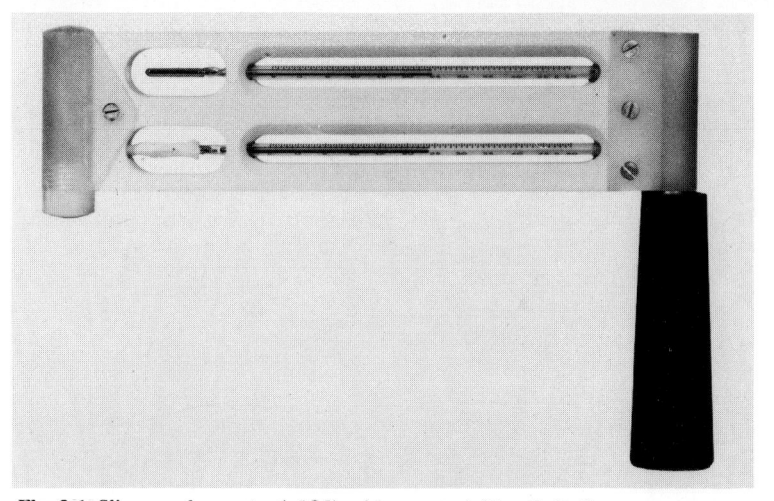

Fig. 3.1. Sling psychrometer (whirling hygrometer) (Casella Ltd).

obtained from the readings of this device either by using a chart supplied with the instrument or by using a psychrometric chart.

Aspirated psychrometer

This is a larger but more precise instrument consisting of a wet and dry bulb thermometer mounted in a frame arranged so that air is induced to flow over the bulbs at a regulated rate by means of a fan powered by a clockwork or electric motor. The wick is not continuously wetted but has to be dipped into a separate container of distilled water before reading is taken. The bulbs of the thermometers are shielded from radiant heat by tubes which act as entry ports guiding the air at a speed of between 3.5 and 5.0m s^{-1} (see fig. 3.2). Relative humidity is obtained as with the sling psychrometer above.

Digital humidity meter

One of the most recent developments in temperature and humidity measurement is the electronic meter which makes use of the wet and dry bulb principle but employs a matched pair of solid state sensors to produce signals for the electronic circuitry to process. A battery powered fan draws air over the sensors one of which is covered by a continuously wetted wick. After about one minute of operation a liquid crystal display indicates dry bulb temperature and relative humidity at the command of a selector switch (fig. 3.3). The humidity sensor is housed in a probe connected to the indicator box by means of a flexible cable. By replacing this with a separate temperature probe, air, liquid and surface temperature can be measured.

Heat

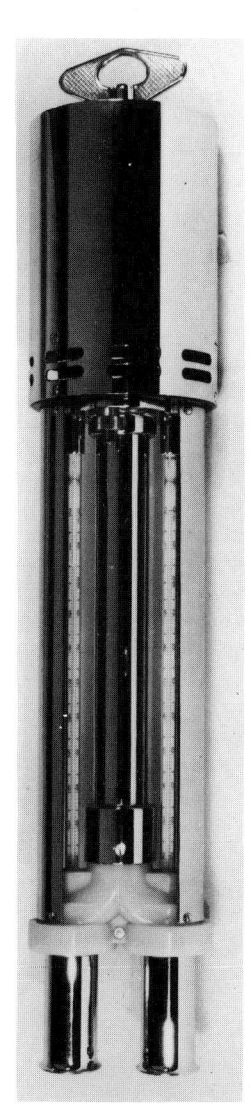

Fig. 3.2. Aspirated psychrometer (Casella Ltd).

Continuous recording of temperature and humidity

The thermohydrograph is the traditional instrument for continuously recording values of temperature and humidity. The temperature is sensed by a bi-metallic element whose curvature changes with temperature. The humidity is sensed by an element consisting of strands of human hair whose length changes with humidity. Both elements are

Chapter 3

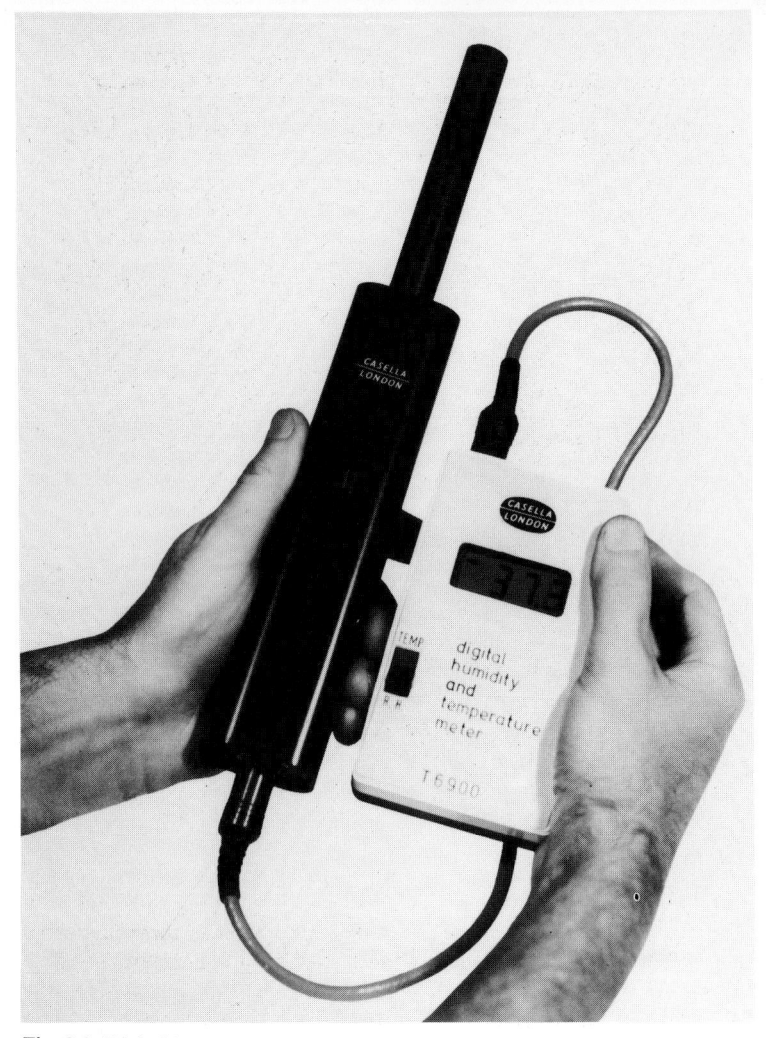

Fig. 3.3. Digital humidity meter (Casella Ltd).

connected via magnifying linkages to recording pens which scribe lines on a paper covered, clockwork driven, rotating drum. These elegant instruments have been popular but functional display pieces for years (see fig. 3.4). The recording period can be either daily, weekly or monthly depending upon the rotating speed of the drum, replaceable paper charts being available for the three periods. With the hair type humidity sensing element there is a loss of precision at the ends of the scale.

For more precise work some of the latest electronic instruments can be coupled to chart or digital recorders.

Heat

Fig. 3.4. Thermohydrograph (Casella Ltd).

Globe thermometer

A mercury in glass thermometer is placed with its bulb in the centre of a matt black sphere or globe. The diameters of the globes available are usually 150mm or 44mm (see fig. 3.5). The larger globe takes approximately 20 minutes to reach equilibrium with its surroundings but the smaller one is quicker.

Kata thermometer

This is an alcohol in glass thermometer having a large silvered bulb at its base and a small bulb at the top of the stem, which is inscribed with a mark top and bottom corresponding to a temperature difference of 3°C. They are available in three ranges: 38° to 35°, 54.5° to 51.5° and 65.5° to 62.5°C. Also inscribed on the stem is a number known as the Kata factor which is specific to each instrument. The device is used to determine the cooling power of the air by timing the rate of fall of the liquid between the two marks having first heated the lower bulb to expand the liquid up the stem. From this the air velocity can be calculated or determined from a chart supplied with the instrument. The Kata thermometer is particularly useful in measuring air velocities below the range of most airflow meters (see fig. 3.6).

Integrating instruments

In order to obtain a single index from wet bulb, dry bulb, globe temperatures and air velocity, instruments are available which measure

Fig. 3.5. Globe thermometers (Casella Ltd).

these parameters electrically and integrate the results into a single value on a meter. This is in addition to having the facility to indicate the individual values in turn by means of a selector switch (see fig. 3.7). As with any instrument that provides a reading on a meter it requires to be calibrated from time to time against accurate mercury in glass thermometers in an environmental chamber.

There is also an instrument known as a 'Botsball thermometer' which consists of a globe thermometer where the complete globe section is covered with a wetted wick. With such a device all four parameters mentioned earlier act upon the globe producing a temperature which approximates to the heat index known as the WBGT (wet bulb globe temperature).

The psychrometric chart

The driving force which makes water evaporate is the difference in 'vapour pressure' between the air and the water surface. The maximum vapour pressure that can occur at any temperature is known as the 'saturation vapour pressure' and this varies with temperature according to the 100 per cent saturation curve on the psychrometric chart shown in fig. 3.8. The vapour pressure and moisture content lines are in the same position on the chart. Curves of relative humidity (percentage saturation) of below 100 per cent are shown lying under this curve in increments of 10 per cent with the dry bulb temperature as the base line and wet bulbs at an oblique angle. Other information shown is moisture content, specific enthalpy and specific volume. This latter unit is the reciprocal of the density of the air and water vapour mixture.

To use this chart it is necessary to measure the wet and dry bulb temperature at a workplace. The point showing the conditions in the workroom air is at the intersection of the lines representing these two values on the chart. All other values can read by extending lines across to the appropriate scales as shown in the example on the sketch in fig. 3.9.

Example

A sling psychrometer measured the wet and dry bulb temperatures to be 10.5°C and 16°C respectively. From the chart in fig. 3.8 the following values can be obtained as indicated by the broken lines in fig. 3.9:

relative humidity (percentage saturation)	50 per cent
moisture content	0.0056kg kg^{-1} (i.e. kg of water vapour per kg of dry air)
specific enthalpy	30.5kJ kg^{-1} of dry air
specific volume	0.826m^3 kg^{-1}
air density (reciprocal of specific vol.)	1.21kg m^{-3}

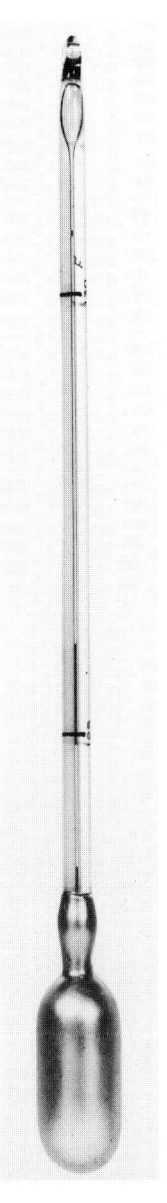

Fig. 3.6. Kata thermometer (Casella Ltd).

Chapter 3

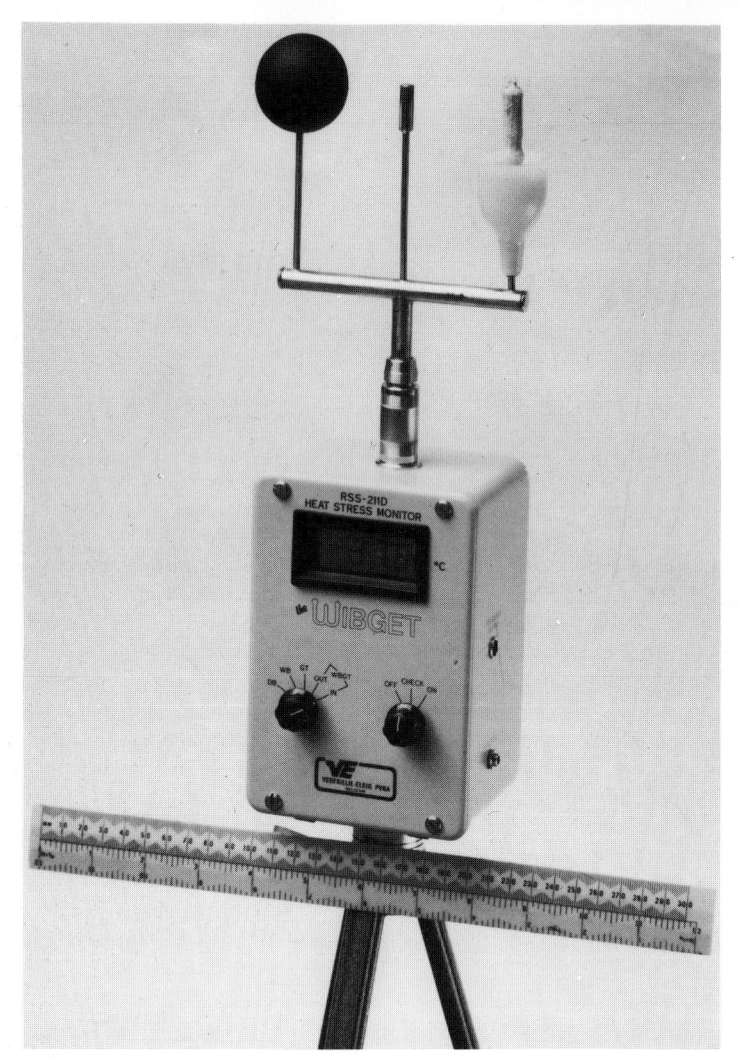

Fig. 3.7. An integrating heat stress meter (Vertec Scientific).

Heat indices

Many scientists have tried to combine all or most of the parameters mentioned previously together with the work rate and clothing worn into a single index which it is hoped would give an indication of the degree of discomfort or stress to be expected of that environment. Many indices have been devised to suit a particular industry emphasising some factors more than others thus may not be ideal for the specific problem under

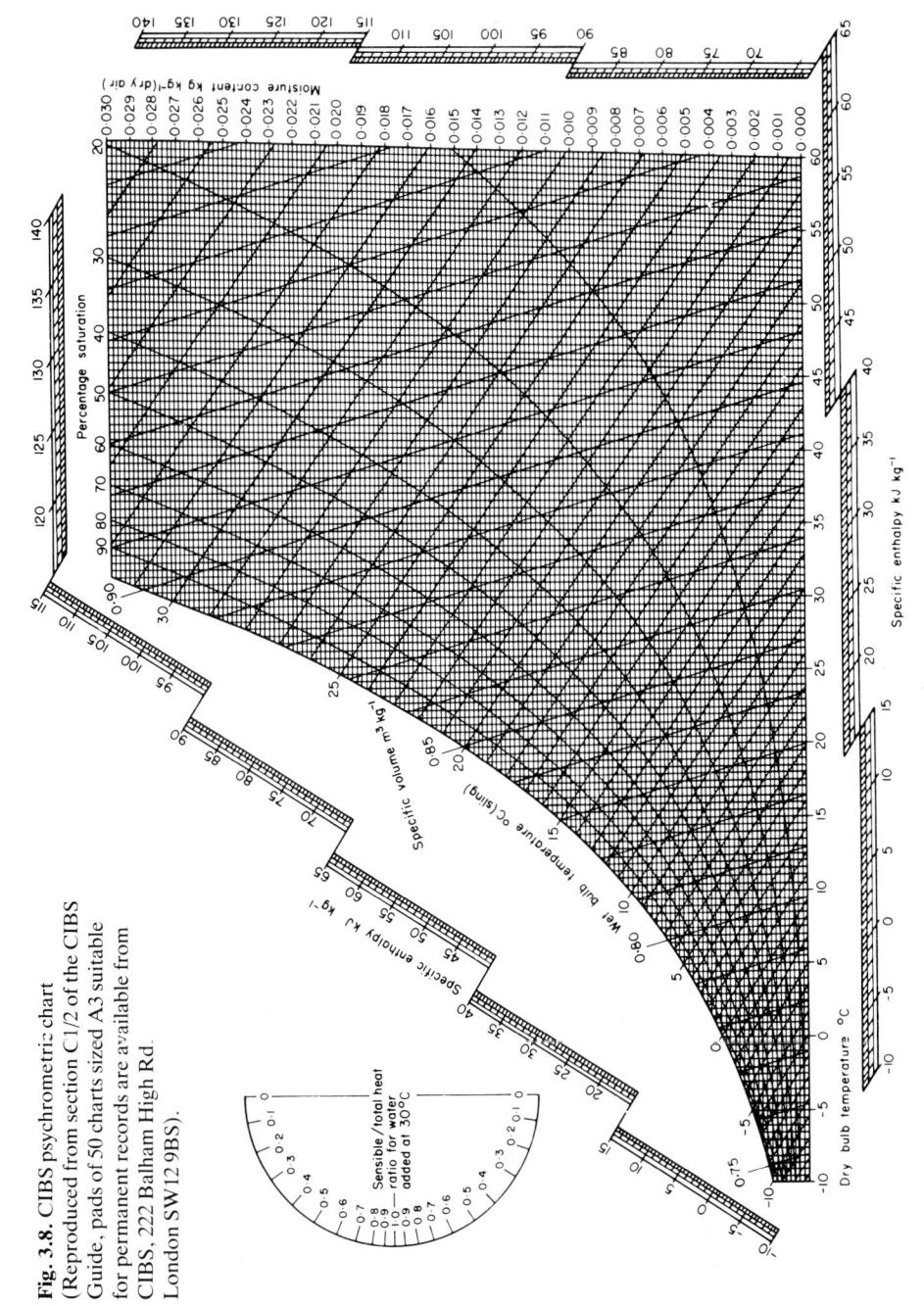

Fig. 3.8. CIBS psychrometric chart (Reproduced from section C1/2 of the CIBS Guide, pads of 50 charts sized A3 suitable for permanent records are available from CIBS, 222 Balham High Rd. London SW12 9BS).

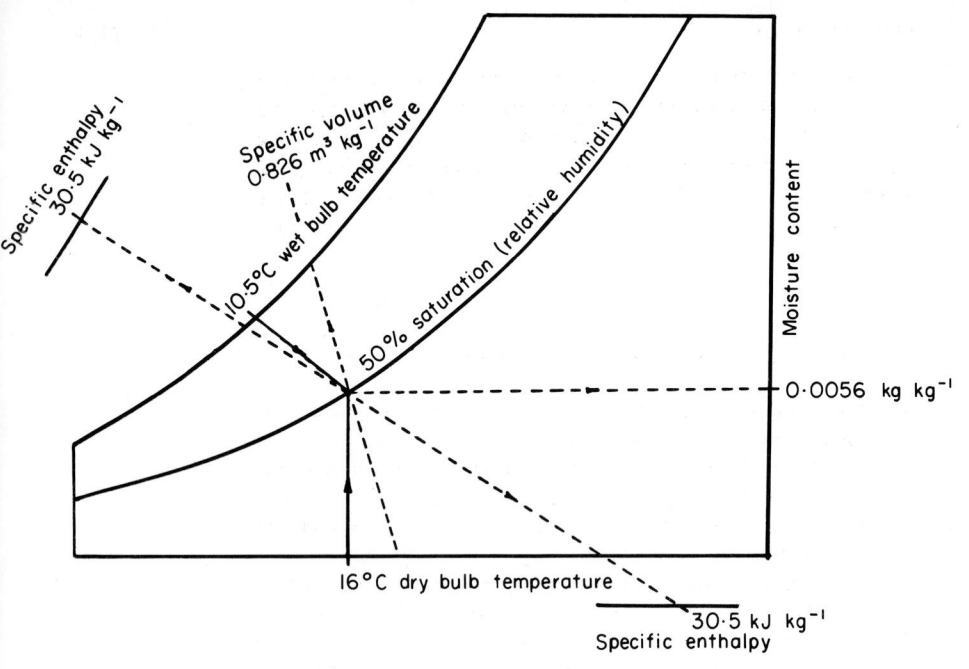

Fig. 3.9. Sketch to illustrate use of psychrometric chart.

consideration. The index which is covered in the American Conference of Government Industrial Hygienists (ACGIH) list of Threshold limit values is the WBGT (wet bulb globe temperature). Given with this index is a table showing maximum values for various work rates together with recommended work and rest periods. An example calculation is given in the following section with table 3.1 showing WBGT values as published by the ACGIH.

Other heat indices can be calculated or obtained from charts and nomograms published in the further reading section given at the end of this chapter.

The measurement of the thermal environment

Aim

It is often necessary to examine the thermal load imposed upon workers in hot or cold industries or in workplaces out of doors in hot or cold climates. By measuring four parameters: wet and dry bulb temperature, air velocity and globe temperature, heat stress indices can be obtained and the thermal components making up the workplace environment can be evaluated so that the more extreme factors can be improved to the benefit of the worker.

Equipment required *Heat*

Kata thermometer, two mercury in glass thermometers of the correct range to suit the environment to be measured, a globe thermometer, stop watch, thermos flask, sling psychrometer, tissues, 25ml beaker, muslin wick, distilled water, rubber bung bored out to take one of the thermometers, aluminium foil, string, scissors, metal polish and soft cloth and a tripod stand with clamps.

Method

1 Before starting the survey fill the thermos flask with very hot or boiling water.
2 Polish the bulb of the Kata thermometer and record the Kata factor inscribed on the stem of the instrument.
3 To prepare the dry bulb thermometer, carefully fit the rubber bung to the lower end and attach a piece of aluminium foil round the bung to shield the bulb of the thermometer but not to restrict any airflow around it. The foil should be fitted shiny side out.
4 To prepare the natural wet bulb, attach the muslin wick over the bulb of the other thermometer covering it completely. Wet the wick with distilled water and allow the loose end of the wick to dip into the beaker containing distilled water. By means of string or sticky tape hang the beaker just below the bulb but clear from it to allow unrestricted airflow.
5 Check that the bulb of the globe thermometer is central in the globe.
6 Arrange these instruments on the tripod stand as shown in fig. 3.10 and place at the workplace to be measured, making sure that the globe is situated where the workers head should be.
7 Allow about 20 minutes for the instruments to reach equilibrium and note and record the values of dry bulb, natural wet bulb and globe temperatures.
8 To measure the air velocity using the Kata thermometer proceed as follows:
> Immerse the bulb of the Kata thermometer in the hot water in the thermos flask and when the alcohol column reaches the upper bulb remove it immediately. Make sure that there is a continuous column of alcohol between the two bulbs as sometimes a vapour lock occurs which disappears if the bulb is further heated. Do not overheat the bulb as the thermometer may burst. Wipe the silvered bulb dry and using a stop watch time the fall of the alcohol column between the two marks on the stem of the thermometer. Repeat the process three or four times noting the times and calculate the mean.

9 Measure the air humidity by using the sling psychrometer. Make sure that the wick is wetted and the reservoir is full of distilled water and that the loose end of the wick reaches the water. Hold the instrument by the

handle and rotate it as fast as possible to allow air to flow over the bulbs for at least one minute. Reading the wet bulb first record both temperatures. This should be repeated several times until the temperatures taken are consistntly the same. Evaluate the relative humidity from the chart provided with the instrument or by using the psychrometric chart. Care must be taken when reading this instrument to ensure that the measurers hands are well away from the bulbs at that time.

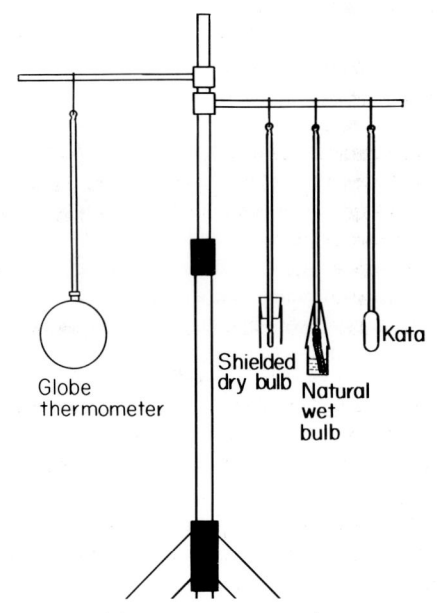

Fig. 3.10. Arrangement of thermometers on stand.

Results and calculations

The results should be recorded as they are taken on a table similar to table 3.2. The air velocity can be calculated from the Kata thermometer results as follows:

$$v = \left\{ \frac{1}{b} \left(\frac{H}{\theta} - a \right) \right\}^2 \quad \text{(from } H = \theta(a+b\sqrt{v}))$$

where: H = cooling power = Kata factor ÷ cooling time in seconds
  θ = mean Kata range — dry bulb temperature
  a & b = constants for the instrument
  v = air velocity

Alternatively use the chart supplied with the instrument an example of which is given in fig. 3.11 with the broken lines representing a specimen example. The mean radiant temperature can be calculated from:

$$(T_r)^4(10^{-9}) = (T_g)^4(10^{-9}) + 0.247\sqrt{v}(t_g - t)$$

where: $T_r = 273 + t_r$
  t_r = mean radiant temperature °C
  $T_g = 273 + t_g$
  t_g = globe temperature reading in °C

Alternatively use the nomogram in fig. 3.12 with the broken lines representing a specimen example.

The WBGT can be calculated from:
for indoor use, $\text{WBGT} = 0.7t'_n + 0.3t_g$
for outdoor use, $\text{WBGT} = 0.7t'_n + 0.2t_g + 0.1t$
where: t'_n = natural wet bulb temperature °C
  t_g = globe temperature °C
  t = dry bulb temperature °C

In order to evaluate this index it is necessary to establish the work rate of the person whose workplace is being mesured, as either; light, moderate or heavy. Some examples are given as follows:

light work rate: sitting, standing with small hand movements
  examples: desk work or light assembly work.

moderate work rate: walking, standing with heavy hand or light arm work,

  examples: supervisory work covering a wide area, messenger work, bench work with heavier items, press operation.

heavy work rate: standing with heavy arm work, work involving whole body,

  examples: sawing and filing, lifting from floor, shovelling, walking carrying loads, pulling or pushing trolleys or barrows, climbing ladders or steep steps most of the time.

Using table 3.1 the recommended maximum WBGT can be established for different work rates.

Table 3.1.

Work/rest regime	Total work load		
	Light	Moderate	Heavy
Continuous work	30.0	26.7	25.0
75% work, 25% rest	30.6	28.0	25.9
50% work, 50% rest	31.4	29.4	27.9
25% work, 75% rest	32.2	31.1	30.0

Other heat indices can be obtained by reference to some of the publications given in the further reading section at the end of this chapter.

Chapter 3   **Possible problems**

1 The Kata bulb may be dirty and this will increase the cooling time and provide an exaggeratedly low air velocity result.
2 Dirty wicks or the use of non-distilled water on the wet bulbs will result in a lowered evaporation rate giving an exaggeratedly high humidity.
3 When reading thermometers make sure that no artificial heat source such as human hands or breath come into contact with them.
4 Ventilated wet bulbs will immediately rise in temperature as soon as the ventilation ceases therefore they must be read quickly to prevent a falsely high humidity being obtained.
5 When undertaking an indoor survey it is useful to note the outdoor conditions with regard to temperature, moisture content and wind velocity as these factors can have an important effect on the inside values.

Table 3.2. Suggested layout for a thermal survey results sheet.

Survey .... Date..................
Location .... Measured by.....................

Measuring station			Readings					Mean
			1	2	3	4	5	
Remarks	time							
	dry bulb (t)							
	natural wet bulb (t'_n)							
	globe (t_g)							
	sling or aspirated	wet bulb (t')						
		dry bulb (t)						
	relative humidity %							
	Kata	factor mean temperature range						
		cooling time (seconds)						
		air velocity (v)						
	mean radiant temperature (t_r)							
	WBGT (wet bulb globe temperature)							

Use of the Kata thermometer chart (fig. 3.11).

Example: Kata factor (cooling factor) 539, temperature range 38°–35°C mean cooling time as measured 95 s, air temperature (dry bulb) 18°C.

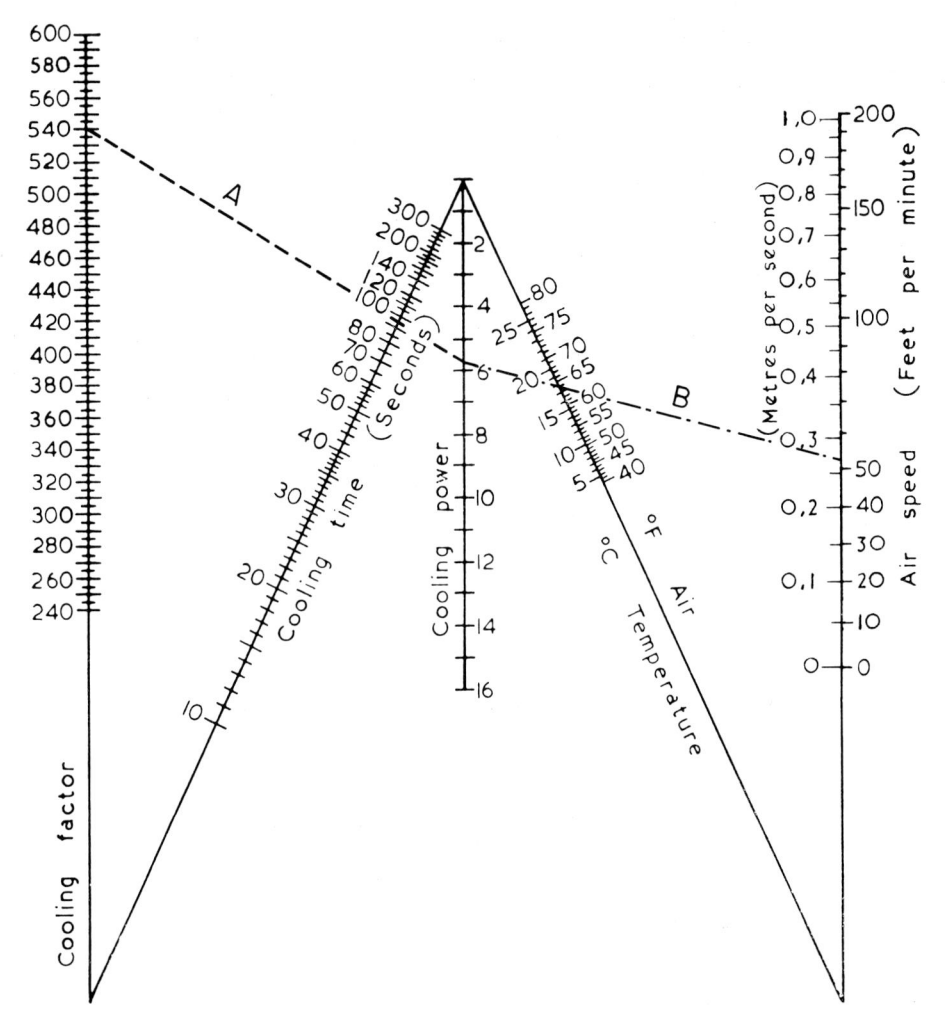

Fig. 3.11. Kata thermometer chart (Reproduced from a withdrawn British Standard, courtesy BSI).

Chapter 3

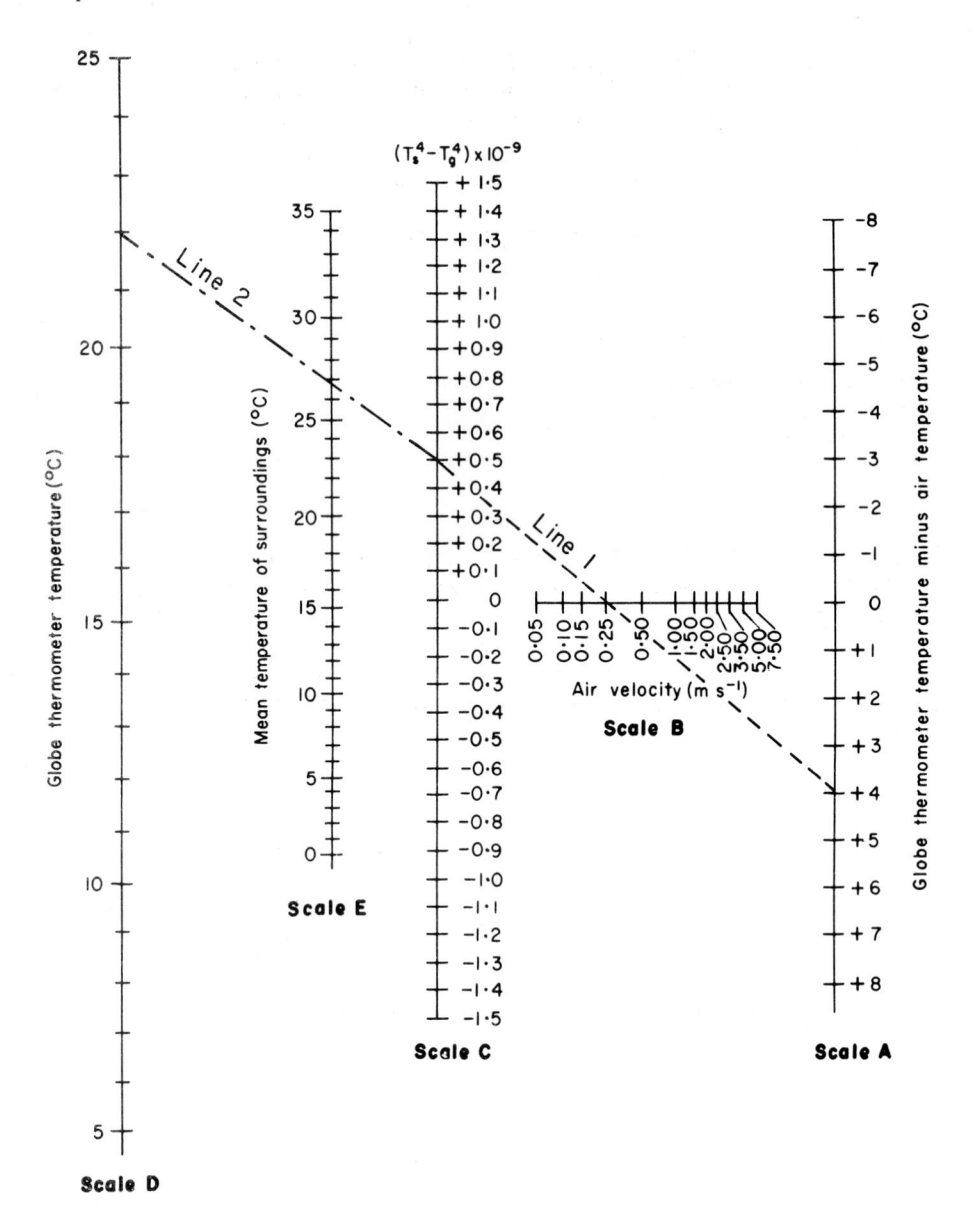

Fig. 3.12. Nomogram for the determination of mean temperature of surroundings (mean radiant temperature) from the globe thermometer (*British Journal of Industrial Medicine*).

Procedure

1   Draw a line (*A*) to join 539 on 'cooling factor' line with the measured mean cooling time 95 s on 'cooling time' line and produce to intersect with the 'cooling power' line (in this example cooling power = 5.7).
2   From that point on 'cooling power' line draw a line (*B*) to join with 18°C on the 'temperature' line and produce it to intersect with the 'air speed' line.
3   Air speed can be read in either m s^{-1} or ft min^{-1} (in this example air velocity is shown to be 0.26 m s^{-1}).

Use of the Globe thermometer chart (fig. 3.12)

Example: globe temperature (t_g) 22°C
air temperature (dry bulb) (t) 18°C
air velocity from Kata (v) 0.26m s^{-1}.

Procedure

1   Subtract $t - t_g$ (22 − 18 = +4) and join $t - t_g$ (+4) on scale *A* to velocity (0.26) on scale *B* and produce to cut scale *C* (line 1). In the example it gives a value of +0.5 on scale *C*.
2   Join that point on scale *C* (+0.5) with t_g (22°C) on scale *D* (line 2).
3   The mean radiant temperature can be read from the point where line 2 crosses scale *E* (mean temperature of the surroundings). In this example it reads 26.8°C.

Further reading

Harrington JM & Gill FS. *Pocket Consultant in Occupational Health*. Grant McIntyre, 1982.
Gill FS. Heat, in Waldron HA & Harrington JM (Eds) *Occupational Hygiene*. Blackwell Scientific Publications, 1980.
Kerslake DMK. *The Stress of Hot Environments*. Cambridge University Press, 1972.
Fanger PO. Thermal comfort, in *Analysis and Application in Environmental Engineering*. McGraw Hill, 1970.

CHAPTER 4
VENTILATION

Introduction

Where natural ventilation does not provide an adequate exchange of fresh air, mechanical devices such as fans may be provided to supply or extract air locally or to ventilate an environment in general. Where the air is required to be moved some distance, ducting is used which may be of considerable length and contain bends, changes of section, branch pieces and other fittings. Coupled with the capacity to draw in fresh air or to recirculate it, the system may contain filters, heaters, coolers, humidifiers or a combination of these and, to prevent atmospheric pollution from the discharge of dirty air, dust collectors and various air cleaners may be used. The performance of these ventilation systems requires to be checked from time to time to ensure their satisfactory operation. This may involve measuring air volume flow rates, velocities and pressures inside ducts and the tracing of airflow patterns around ventilation terminals such as: extraction hoods, slots, enclosures and fume cupboards. The routine checking of ventilation systems is mandatory in places required to be ventilated under the Asbestos Regulations and it is recommended for all other places where it has been considered necessary to maintain comfort or a healthy working environment.

Natural ventilation is difficult to measure as it relies upon pressure differences created by natural forces such as the wind or differences in air temperature between inside and outside the building and therefore is constantly fluctuating unlike the more steady mechanically induced airflows. Nevertheless natural airflows can be estimated using a technique outline in this chapter. Air volume flow rates are quoted in units of cubic metres per second ($m^3 \, s^{-1}$) or in air changes per hour. For natural ventilation rates air changes per hour is more often used. This unit can be converted to $m^3 \, s^{-1}$ by multiplying by the volume of the room in cubic metres and dividing the result by 3600. Air speeds are quoted in metres per second ($m \, s^{-1}$).

Air pressures are usually quoted as gauge pressures, that is, the pressure difference between inside the system and atmospheric pressure or that of the room in which the equipment is installed. In SI units pressure is quoted in the unit Newton per square metre ($N \, m^{-2}$) which is also known as a Pascal, (Pa), that is, $1 Pa = 1 N \, m^{-2}$. However, as pressure gauges are often simple U-tubes containing a liquid such as water or paraffin, tradition has it that pressures are sometimes quoted as the length or height of a column of liquid, for example, inches or millimetres of water. In order to convert a column of liquid to Pascals the following equation is used:

$$\text{Pressure, } p = \rho'gh \text{ Pa}$$

where: ρ' = the density of the liquid in the gauge in kg m^{-3}
h = the height of the column in metres
g = the acceleration due to gravity (9.81m s^{-2})

Example: A column of water of height 100mm expressed as a unit of pressure, Pa, is calculated knowing that the density of water is 1000kg m^{-3} from

$$p = 1000 \times 9.81 \times \frac{100}{1000} = 981 \text{Pa}$$

Some further explanation of pressure is necessary in order to differentiate between static pressure, velocity pressure and total pressure in ventilation stems.

Static pressure is the pressure exerted in all directions by a fluid that is stationary. If it is in motion it is the pressure exerted at right angles to the direction of flow. It can be either positive or negative for example, on the suction side of a fan it would be negative but on the delivery side positive in relation to atmospheric pressure.

Velocity pressure is defined as the pressure equivalent of the kinetic energy of a fluid in motion but it is best illustrated as that pressure which is exerted on a surface placed across an airstream as in the sail of a boat or the vanes of a windmill thus causing them to move if sufficiently strong. It is calculated from the expression:

$$\text{velocity pressure, } p_v = \rho \frac{v^2}{2} \text{ Pa}$$

where: ρ = the density of the air in kg m^{-3}.
v = the velocity of the air in m s^{-1}.
Velocity pressure is always positive.

Total pressure is the sum of the static and the velocity pressure at a point in an airstream and can be either positive or negative in relation to atmospheric pressure.

Equipment available

Pressure measuring instruments

It is possible to make a simple pressure gauge or 'manometer' by taking a glass tube bent in a U-shape and filling it half full with water. With one limb connected to the inside of the ventilation duct by means of rubber

Chapter 4

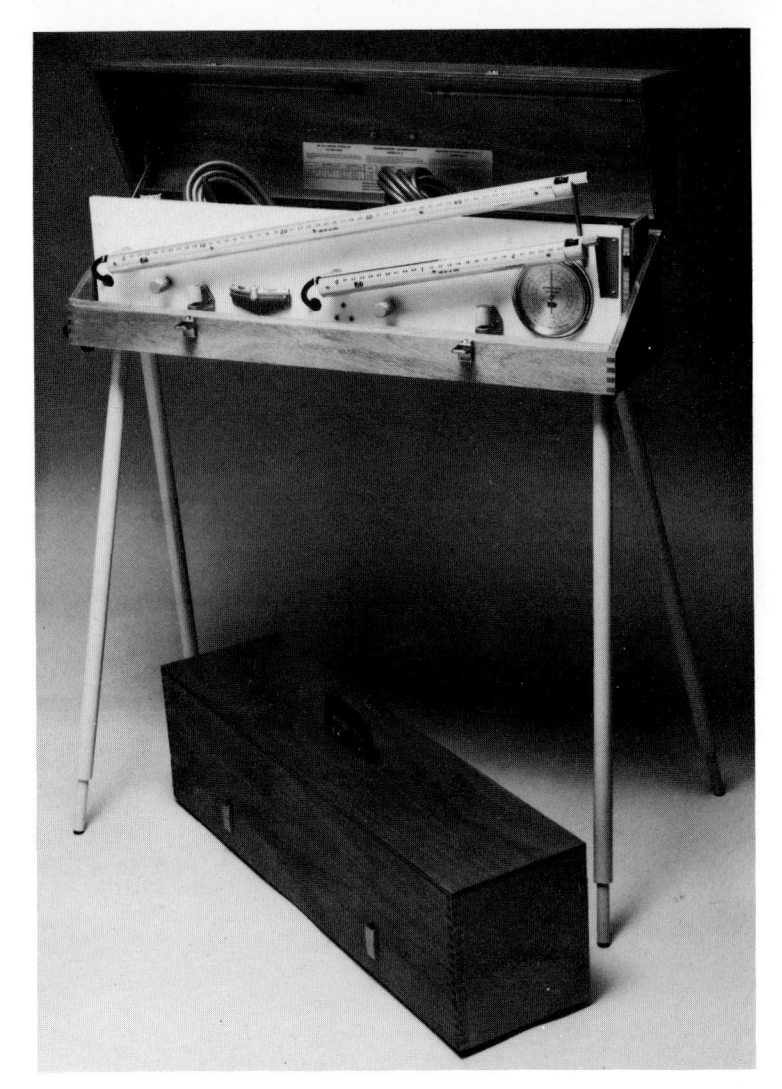

Fig. 4.1. Portable inclined manometer (Airflow Developments Ltd).

or plastic tubing at the point where the pressure is required and with the other limb open to the atmosphere the water will take up different levels in each limb. The difference in vertical height between the two liquid levels represents the pressure in the duct. Using the formula given above this pressure can be calculated. For very low pressures this home made instrument will be imprecise because it will be difficult to measure the difference in height of the two columns. If more precision is required then it is necessary to incline one limb of the U-tube as in fig. 4.1 and the length of the liquid in the inclined tube can be measured and the difference in height of the two limbs then calculated knowing the angle of inclination of the inclined tube.

There is a wide range of commercially available manometers from simple vertical U-tubes to instruments whose angle of inclination can be varied in known fixed positions and whose scales are calibrated in various units of pressure. The liquids used in the more sophisticated instruments are usually of a lower specific gravity than water thus providing a more extended scale than with water for the same pressure. It is important to ensure that such gauges are filled with the correct liquid or the scales will read the incorrect value. Also with gauges having variable inclinations it is important to multiply the reading obtained by a scale factor appropriate to the angle of inclination as provided by the maker.

Liquid filled gauges have the disadvantage that the liquid must be free of bubbles when being used and overloading the gauge can result in bubbles being formed or the liquid being blown out of the tubes. Also vertical U-tubes must be held vertical when in use and inclined manometers must be carefully levelled and zeroed before use and kept level whilst being read.

Diaphragm gauges do not have these disadvantages as a reading is obtained on a dial by a pointer actuated by the movement of a diaphragm one side of which is exposed to the pressure to be measured. A mechanical or magnetic linkage moves the pointer. Such gauges are much easier to use in industrial situations but they require to be calibrated from time to time against an accurate inclined manometer set up

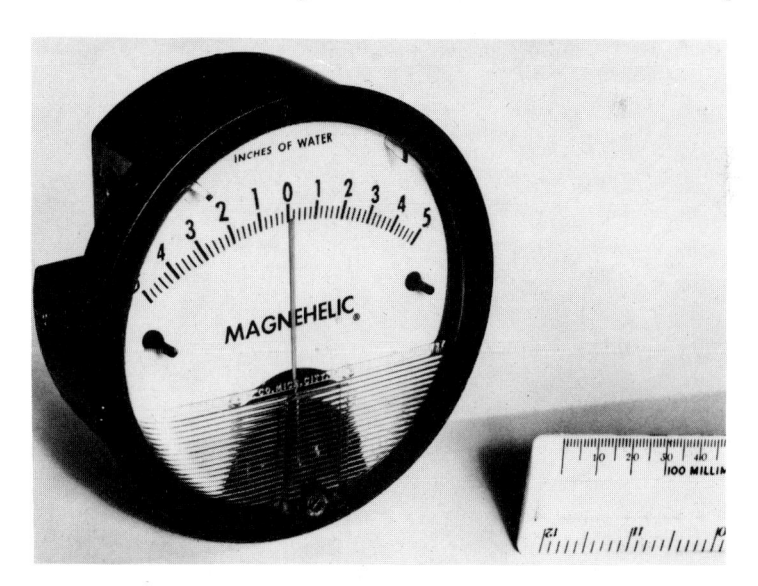

Fig. 4.2. Diaphragm pressure gauge (Granville Controls).

Chapter 4

in the comparative quietness of a laboratory. They also require to be zeroed before taking a set of readings which can be done by joining the two pressure tappings by a tube and making the appropriate adjustment. Gauges of this type can be read placed either in a vertical or horizontal position but they must be zeroed in the plane in which they are to be read. An example of a diaphragm gauge is shown in fig. 4.2.

Air velocity measuring instruments

There is a wide variety of air velocity instruments available and they can be classified into three main groups: vane anemometers, heated head anemometers and velocity pressure devices.

Vane anemometers

These are small rotating windmills mechanically or electrically coupled to a meter or a digital indicator. The mechanically coupled ones (fig. 4.3) require to be used in conjunction with a timing device such as a stopwatch or electronic timer but the electric vane anemometers (fig. 4.4) give a direct reading in units of air velocity such as metres per second (m s^{-1}). The modern instruments are available in diameters varying from 100mm to 25mm but as the rotating vanes are extremely light and are suspended on jewelled bearings it is important to ensure that they are handled carefully and nothing allowed to touch the vanes. This type of

Fig. 4.3. Mechanical vane anemometer (Airflow Developments Ltd).

Ventilation

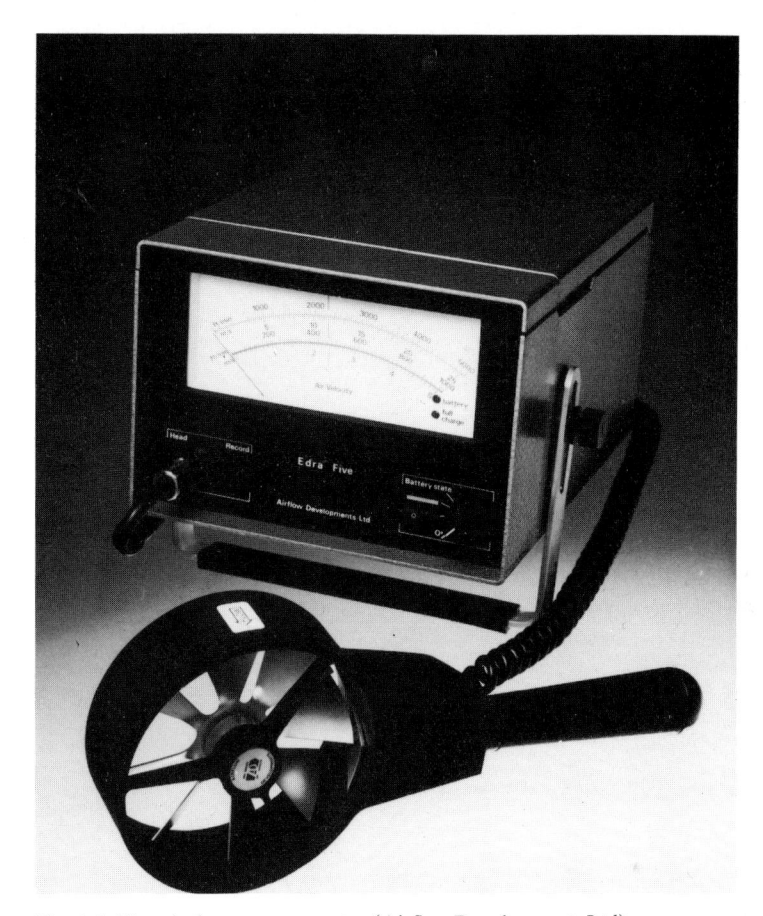

Fig. 4.4. Electrical vane anemometer (Airflow Developments Ltd).

instrument requires calibrating regularly if reliable results are to be obtained. The electrically powered instruments should not be used in flammable atmospheres unless they are certified intrinsically safe.

When using these instruments it must be borne in mind that they are sensitive to 'yaw' that is, their axis must be parallel to the stream lines.

Heated head air meters

These devices rely upon the cooling power of the air to cool a sensitive head. Although the Kata thermometer mentioned in the chapter on heat measurement could be classified as one of these it is not described here. Essentially the heated head is a hot wire, thermocouple or thermister bead through which an electric current is passing to maintain it at a constant temperature. As air blows over it cooling takes place depending

Chapter 4

upon the air velocity. The current which is required to keep the temperature constant is registered on a meter which has been previously calibrated in units of air velocity. An example of this is shown in fig. 4.5. As with vane anemometers they require careful handling and regular calibration against known air speeds. Also they may not be used in flammable atmospheres.

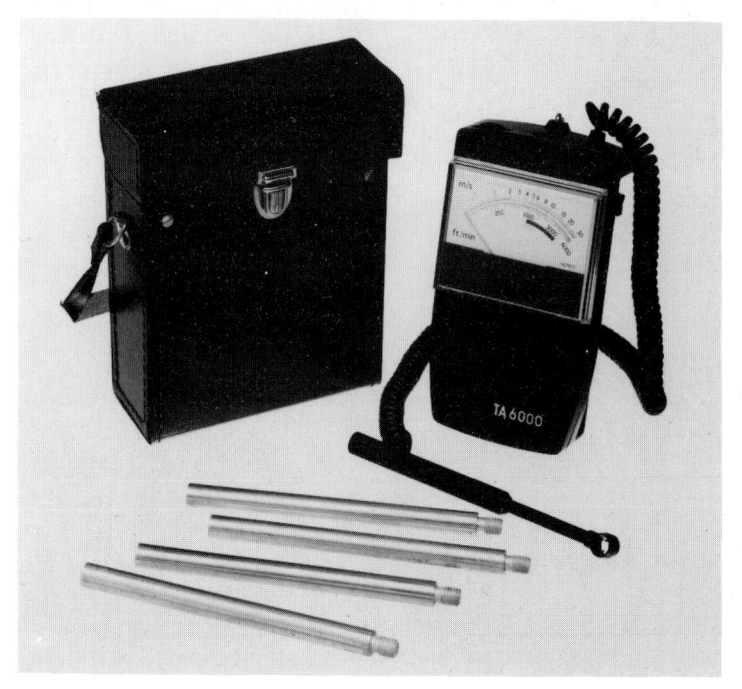

Fig. 4.5. Thermal anemometer (Airflow Developments Ltd).

Some of these instruments have a cowl over the sensing head to direct air over it which means that they must be carefully placed in the airstream with no yaw.

Velocity pressure devices

It is possible to measure velocity pressure using a pressure gauge as described above in conjunction with a probe known as a 'pitot-static' tube. This device consists of two tubes, one concentrically placed inside the other. The inner tube is positioned facing into the airstream with its axis parallel to the stream lines sensing the total pressure. The outer tube is sealed at the end with aerofoil shaped seal allowing only a small opening for the inner tube. Around the outer tube is a ring of holes which are at right angles to the airstream sensing static pressure. At the opposite end to the sensing head tubes are fitted each with a nozzle to connect to the pressure gauge via flexible tubing. In order to facilitate insertion into the side of ducting the whole device is bent at a right angle. The principle of operation is as follows: as the static pressure inside the duct is acting upon both tubes it is also acting upon each side of the pressure gauge and therefore cancels itself out leaving only the velocity pressure to provide the reading on the gauge. This is illustrated in fig. 4.6. Rearranging the velocity pressure formula mentioned earlier the air

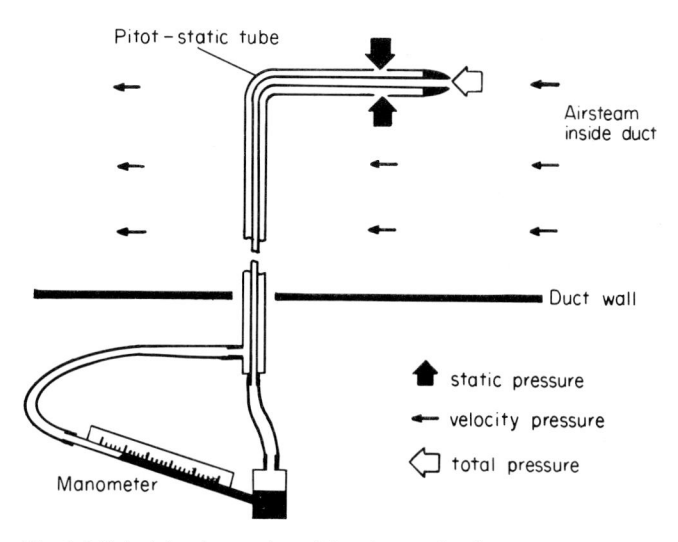

Fig. 4.6. Principle of operation of the pitot-static tube.

velocity can be calculated from the measured pressure as follows:

$$\text{velocity}, v = \sqrt{\frac{2p_v}{\rho}} \text{ m s}^{-1}$$

Example: a velocity pressure of 100Pa is measured using a pitot-static tube, assuming air density is 1.2kg m^{-3} the air velocity is:

$$v = \sqrt{\frac{2 \times 100}{1.2}} = 12.9 \text{m s}^{-1}$$

It is important to note that this instrument is sensitive to air velocities above 3m s^{-1} but below this precision is lost and an alternative airflow meter should be used. Also with pitot-static tubes it is important to ensure that the tube points directly into the airstream, any deviation will result in errors, to this end some makes of pitot-static tube have a pointer at the lower end to indicate the direction in which the head is pointing.

Chapter 4

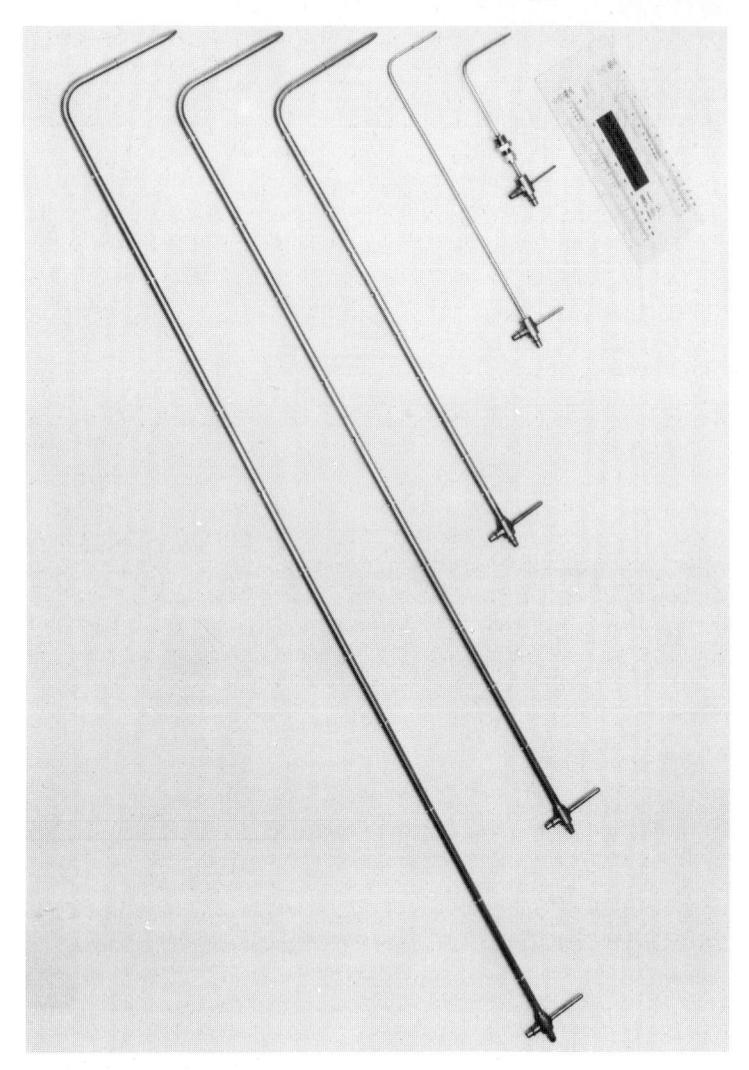

Fig. 4.7. Pitot-static tubes (Airflow Developments Ltd).

Smoke tube kit

To trace the patterns of airflow outside ducts, to spot leaks of air and to identify source of draught, a smoke tracer is available. White smoke can be produced in a variety of ways but the most convenient is to purchase a smoke tube kit which consists of a rubber bulb fitted with one way valves and a small sealed glass tube containing a chemical which when exposed to air gives off clouds of white smoke. To activate the chemical it is necessary to break the sealed ends of the glass tube, attach one end to the bulb. By squeezing the bulb air is passed through the tube producing a stream of white smoke. Rubber end seals can be fitted to the tube to conserve the chemical for later use. There is a tendency for the tube to seal itself with a deposit of white powder before all the chemical is exhausted but this can be cleared with a thin spike.

A useful feature of smoke produced this way is that it is at the same temperature as the surrounding air. It is not recommended that cigarette smoke be used as it is hotter than the surrounding air and will tend to rise giving a false picture of the airflow patterns.

Calibration devices

It is necessary to check the performance of air velocity instruments from time to time against accurately known air speeds. This is best done in a wind tunnel but such devices are large, expensive and are limited in location to the larger research establishments, universities and polytechnics and may not be accessible to all. However there is a small, relatively inexpensive wind tunnel commercially available that is known as an 'open jet' wind tunnel and can guarantee air velocities to within ±2 per cent of the true value. This is sufficiently precise for most field measurements. The device can be installed on an open bench provided there is at least 6m of unrestricted horizontal space available. A description of the method of use of this tunnel is given later.

Barometric pressure instruments

Barometric pressure can be measured by a variety of devices. The absolute instrument is the Fortin barometer which is a column of mercury contained in an inverted sealed tube standing in a trough of mercury. The column is held up by air pressure whose value is expressed by the height of the column supported in mm, for example, standard atmospheric pressure is 760mm of mercury (Hg). This value can be converted into Pascal and hence into bar or millibar (mb) by using the formula in page 71 (1bar = 10^5Pa or 1mb = 100Pa). This pressure is also measured by an aneroid barometer which is essentially a sealed bellows type chamber which expands and contracts with changes in pressure and moves a needle on a suitably calibrated dial. The latest devices use pressure transducers which can provide a signal to a digital display.

The measurement of airflow in ducts

Purpose

Ducted ventilation systems require checking at regular intervals to

Chapter 4

ensure that the designed airflow rate is being maintained. Deterioration occurs gradually due to a variety of factors, but mainly because of a build up of deposits on the fan blades, duct walls and other parts of the system. If a record is made of the results routinely, then trends can be spotted and remedial action taken.

Equipment required

An air speed measuring device which can be either a vane anemometer, a heated head air velocity meter or a pitot-static tube and pressure gauge; a tape measure and a marker pen. It should be borne in mind that whatever instrument is chosen, an access hole or holes are required in the duct wall to allow the instrument to be inserted, thus the smaller the instrument the smaller the hole. In this respect, the pitot-static tube is the most suitable requiring a hole no larger than about 12mm in diameter, although some of the smaller heated head instruments will also fit that size hole. Vane anemometers are less suitable for this purpose.

Note that if the ducting handles air containing radioactive, pathogenic material or toxic chemicals, then drilling holes may lead to contamination of the drill, the measuring instruments and escape of the pollutant into an occupied area.

Method

1 Select a suitable length of duct in the airstream to be measured. Ideally the measuring station should be in an airstream free from turbulence which means the ducting should be straight and there should be no obstruction or changes of direction in the duct for at least 10 diameters upstream of it and none should appear downstream for 5 diameters. In many installations it is not possible to find such a place, therefore the longest length of straight ducting should be chosen and the measuring station taken as far downstream as possible from the last cause of turbulence. Unfortunately, the reliability of the results will be affected by the degree of turbulence in the airstream.

2 Having chosen a measuring station it is necessary to place the instrument's sensing head in representative places over the cross-section of the duct in order to obtain an average velocity. This is because the air moves at a higher speed towards the centre than it does close to the walls of the duct. The British Standard BS 848 recommends positions for the instrument as set out in figs 4.8 and 4.9 for circular and rectangular cross-sections. Circular ducts are divided according to a Log-Linear Rule and the rectangular according to a Log Tchebycheff Rule. Thus it may be necessary to drill various holes in the duct walls to suit the size of the measuring station. Plugs of rubber or other suitable material should be available to cover the holes after use.

Ventilation

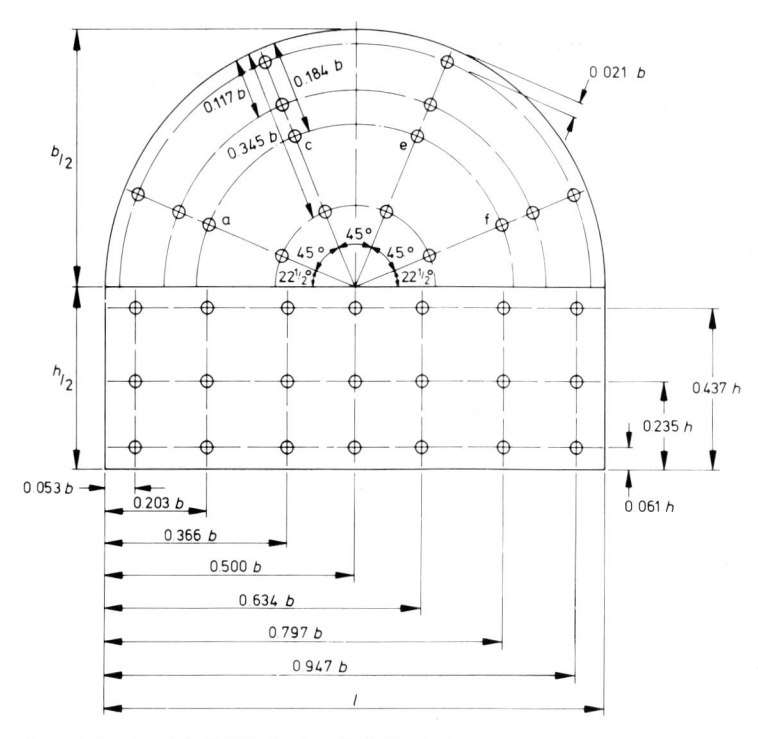

Rectangular base: lower half of 7 × 6 log/Tchebycheff rule (21 points)
Semicircular area: log/linear rule. 4 radii, 4 points per radius (based on 4 diameters, 8 points per diameter) (16 points)
Total number of measurement points: 37

Fig. 4.8. Log-linear rule for traverse points on 3 diameters in a circular duct. (Extract from BS 848: Part 1: 1980 reproduced by permission of the British Standards Institution, 2 Park Street, London W1A 2BS from whom complete copies of the standards can be obtained).

3 In order to assist in placing the sensing head of the instrument in the correct position inside the duct, the stem or carrying arm should be marked so that when the mark is aligned with the side of the duct the head is in one of the positions indicated by figs 4.8 or 4.9. For example, if the duct is circular in cross-section and of 300mm diameter, the stem should be marked at the following eight distances from the head:

$300 \times 0.021 = 6$mm $\quad 300 - 104 = 196$mm
$300 \times 0.117 = 35$mm $\quad 300 - 55 = 245$mm
$300 \times 0.184 = 55$mm $\quad 300 - 35 = 265$mm
$300 \times 0.345 = 104$mm $\quad 300 - 6 = 294$mm

4 With battery driven instruments, check the battery as indicated by the maker and turn on the meter and, where necessary, zero the scale in still air. This can be done by placing the head in a small closed container such as a tube sealed at one end.

Chapter 4

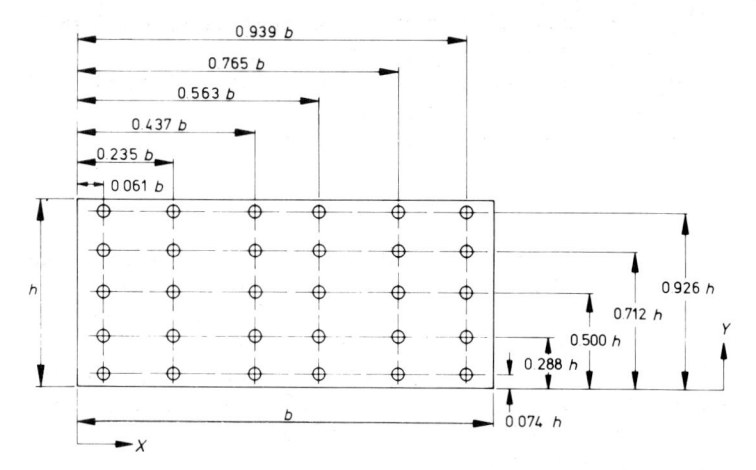

Fig. 4.9. Log Tchebycheff rule traverse points in a rectangular duct (Extract from BS 848: Part 1: 1980 reproduced by permission of the British Standards Institution, 2 Park Street, London W1A 2BS from whom complete copies of the standards can be obtained).

5   With pitot-static tubes it is necessary to set up the pressure gauge. Inclined manometers require to be levelled and zeroed, but this is best done after the connecting tubes have been attached so as not to disturb the setting by so doing. The high pressure side of the gauge must be connected to the central tube of the pitot-static tube and the low pressure side to the outside tube. Diaphragm gauges must be zeroed in the position in which they are to be read, i.e. either flat or vertical, but the position must not be changed after that.

6   Remove any protective cover from the sensing head and insert it through the hole in the duct and position it so that the first mark is against the duct wall. If the duct is circular, then the stem of the instrument must be held along the line of the diameter, but if it is a rectangular cross-section then the stem must be at right angles to the duct wall. Ensure that the sensing head is exactly facing into the airstream. This is important with instruments which have a directional shield around the head and with pitot-static tubes which often have a pointer at the lower end to indicate the position in which the head is facing. Some heated head instruments have no directional characteristics so this is not important. When using pitot-static tubes and liquid filled manometers it is a wise precaution to bend and squeeze both flexible connecting tubes to cut off the pressure in the gauge, releasing them simultaneously only when the pitot-tube is in the correct position. This can prevent an excessively high static pressure from blowing the fluid out of the gauge. If the air temperature and barometric pressure are very different from standard conditions (20°C and 1013mb), then they should also be measured.

7 Note the reading of the meter or the gauge and convert it to Pa according to the scale factor or the conversion factor.
8 Move the head to the next position and repeat 7 and so on until all positions in the diagram have been measured.
9 It is wise to repeat the complete set of readings at least once more.

Calculations

Where air velocity meters have been used, it is necessary to correct each reading from the latest calibration chart for the instrument concerned. The corrected velocities in each set of results should then be averaged to obtain the average air velocity at the measuring station.

The pitot-static tube and gauge method has obtained a set of velocity pressure readings which must be converted to a velocity before calculating an average. *Do not average the pressure readings.*

Calculate the air velocity as follows:

$$\text{velocity}, v = \sqrt{\frac{2p_v}{\rho}} \text{ m s}^{-1}$$

where p_v = the reading of the velocity pressure in Pa.

ρ = the air density in kg m^{-3} which can be taken as 1.2kg m^{-3} unless the air temperature or barometric pressure is very different from 20° or 1013mb.

To correct for temperature and pressure:

$$\text{new density } \rho_1 = \frac{1.2 \times P_{at} \times 293}{1013 \times (273 + t)} \text{kg m}^{-3}$$

use ρ_1 in place of ρ in the above formula
where P_{at} = barometric pressure in millibars (mb)
t = duct air temperature in °C

The volume of air flowing is calculated from the formula:

$$\text{Volume flow rate}, Q = \bar{v}A \text{ m}^3 \text{ s}^{-1}$$

where $\bar{v}$ = the average air velocity at the measuring station in m s^{-1}
and A = the cross-sectional area of the duct at the measuring station in m²

With circular cross-section

$$A = \frac{\pi d^2}{4} \text{ m}^2$$

where d is diameter of duct in metres
and with rectangular cross-section $A = a \times b$ m²
where a and b are the length and breadth of the rectangular duct.

Chapter 4   Possible Problems

1 Ventilation flow rates fluctuate for a variety of reasons:
(a) some fans have a fluctuating output;
(b) external influences such as wind and weather can affect flow rates;
(c) internal influences such as the movement of large loads across the entrance or exit of a ventilation system can affect flow rates.
Therefore, it is important to take more than one set of readings.
2 Some flow rates pulsate, so when reading meters, some estimate of the mid-point of the pulse should be taken.
3 Due to deposits of dust or sticky particles on the insides of ducting, the cross-sectional area may not be as imagined. This is particularly prevalent in the ducting on paint spray booths. Adjustments can only be made by inspection.

The measurement of pressure in ventilation systems

Purpose

In order that air should flow in ventilation systems it is necessary to create a pressure difference between the inside of the duct and atmosphere. This is usually achieved by means of a fan. If the ducting is connected to the suction side of the fan the pressure inside is negative thus drawing air into the system. A positive pressure is found on the discharge side, thus ducting on that side will deliver air under the influence of that pressure. Pressure is absorbed by the ducting, fittings and obstructions such as dampers and filters. With some ventilation systems dirt can build up inside which can restrict the flow and increase the pressure absorbed. Also filters will gradually increase in resistance as dust is collected, resulting in an increase in pressure absorbed. It is, therefore, useful to measure pressures at various places in the system and if done routinely, can provide an indication of any deterioration in performance.

Equipment required

A manometer or diaphragm gauges, flexible plastic tubing of sufficient length suit the siting of the gauge in relation to the measuring point (if pressure is to be measured on either side of an item, then it is useful to have tubing of two different colours), some moulding clay or plasticine to act as a sealer, a hand or electric drill capable of drilling holes the same diameter as the outside diameter of the flexible tubing, some rubber plugs to seal up the hole after measuring.

Method

1 Select places to insert the tubing to suit the requirements of the system. For example, on either side of a filter, on either side of the fan,

and/or at chosen places along the ducting. Drill a hole into the duct at each place. If the site is to be measured regularly then it may be advantageous and labour saving if a permanent nozzle is soldered or fixed to the duct wall at the hole to take the flexible tubing. The inner edge of the nozzle must be flush with the inside of the duct so that none protrudes into the airstream, and a cap should be fitted to prevent leakage after use.

Note that if the ducting handles air containing radioactive, pathogenic material or toxic chemicals, then drilling holes may lead to contamination of the drill, the measuring instruments and escape of the pollutant into an occupied area.

2   Connect a length of flexible plastic tubing to each side of the gauge.
3   Level and zero the gauge.
4   Using the highest pressure range available, connect the other ends of the tubing to the places to be measured. With liquid filled manometers it is wise to carefully think about the pressures to be measured before connecting the tubing into the system as liquid can easily be removed by the ventilation pressure. The following points should be memorised:
(a) positive pressure will depress and negative pressure elevate the liquid in the limb to which it is connected,
(b) if a gauge pressure is to be measured, that is, with one side of the gauge open to atmosphere, then all parts of the ventilation system on the suction side of the fan will be at a negative pressure, whilst those on the delivery side will be at a positive pressure,
(c) if a pressure difference is to be measured between two parts of the system, that is, both tubes are connected into the ducting, then with the exception of the fans themselves, the air flows from the high to the low pressure which means that the upstream side will be at the higher pressure,
(d) with fans the higher pressure will be on the delivery side.
5   Having selected the highest range, if the pressure reading is low then a lower range can be used.
6   Note the reading, remove the tubing and seal the hole.
7   The recorded value must be multiplied by the scale factor of the range used.

Results

Little information can be gained from the results unless the design values of the system are known against which the measurements can be compared. However, if routine measurements are taken, say once every month, and a continuous record kep then the condition of the ventilation system can be observed regularly and if any deterioration is noted then corrective action can be taken. It is suggested that for each measuring station a record sheet be kept containing the information as shown in table 4.1.

Table 4.1. Suggested ventilation record sheet, one page per measuring station.

Measuring point ....

Date	Time	Pressure Pa	Airflow rate $m^3 s^{-1}$	Remarks	Instrument used	Measurer's name

Possible problems

1 Ventilation pressures may fluctuate due to external influences. If this is the case, then the cause should be removed if possible and the readings repeated.

2 If the pressure continues to fluctuate or pulsate then attempt to read a mean or mid-point of the pulsation.

3 If the ventilation pressure has been misjudged when using a liquid filled gauge and the liquid has bubbled or blown out, then it is important to ensure that the flexible tubing is cleared of fluid and the liquid replenished in the gauge and all bubbles removed. It may be necessary to hang the tubing in a vertical position to drain for some time or to blow it through with a jet of compressed air. Bubbles may be removed by gently rocking the fluid in the gauge from side to side using a blowing action through a short length of clean tubing.

To measure the performance of a suction inlet

Purpose

Extract hoods, slots, enclosures and fume cupboards are intended to capture air pollutants to prevent them from being released into the general room atmosphere. Unfortunately many have insufficient air flowing or have their suction inlet too far away from the point of release of the pollutant or they have inadequate enclosure around the source.

Thus some means of checking the airflow patterns and air velocity in and around the inlet is useful so that the full extent of the zone of influence of the device can be ascertained.

Equipment required

Smoke tube kit, a tape measure and an air speed indicator such as a thermister bead flow meter or a vane anemometer (the former is to be preferred as it will record at lower air speeds and is less sensitive to air direction). Ideally a thermister bead instrument with an unshielded head would be best.

Method

It should be pointed out at this stage that this method is intended to trace airflow patterns and measure air velocities around the inlet and not to check on the absolute efficiency of the suction device. To achieve the latter it would be necessary to release a tracer gas whose decay of airborne concentration with distance from the source could be measured using some direct reading analysis instrument.

1 Ensure that cross draughts and local air turbulence is minimised during the test.
2 Break the ends of the smoke tube and insert it in the rubber bulb, puff smoke around the suction inlet gradually moving further away from the mouth until the full extent of the zone of influence can be observed. The places where the smoke is no longer being drawn in marks the edge of this zone.
3 Make a sketch drawing of the inlet and the equipment of the workplace that lies within the zone of influence.
4 Plot on the sketch an imaginary grid of squares across the face of the inlet and in the area in front to cover the whole zone of influence in the horizontal plane containing the source of pollutant. Other planes both vertical and horizontal can be chosen if a full picture is required. The dimensions of the grid squares should be 100 to 150mm depending upon the size of the inlet, the smaller ones using the smaller squares.
5 Using the tape measure as a guide to the measuring positions place the sensing head at the corner of each grid square taking care to ensure that the instrument is, as far as possible, axial to the airstreams lines. This may be difficult as the airstreams around suction inlets are curved as air enters from all sides. Also the sensing head should be carried on a long probe so that the position of the observer's arms and body do not interfere with the flow patterns. Note the air velocity at each place by writing it on the sketch at the appropriate position (see fig. 4.10).
6 Measure the air velocity on the face of the inlet along the centreline at the intersection with each of the grid lines. Note the results on the sketch.

Ventilation

Chapter 4

7 Repeat item 5 above for as many planes as have been chosen.

Results

Correct each reading using the calibration chart for the instrument used. Make an average of the velocities across the face of the inlet and note the range of the readings in relation to the average. If there is a wide variation then the airflow distribution is uneven and may require some means of equalisation using face slots or airflow splitters or guides.

Inspect the results on any chosen plane of grids and try to estimate the position of points of equal velocity, for example, draw a cross or spot at the points where in your opinion, the 10, 5, 1, 0.5 and 0.25m s^{-1} velocities occur. Join up the spots of equal velocity to produce a velocity contour of that air speed as shown in fig. 4.10. The lowest contour should be 0.25m s^{-1}. Any pollutant released outside this contour is unlikely to be captured or drawn into the inlet even if released into an undisturbed

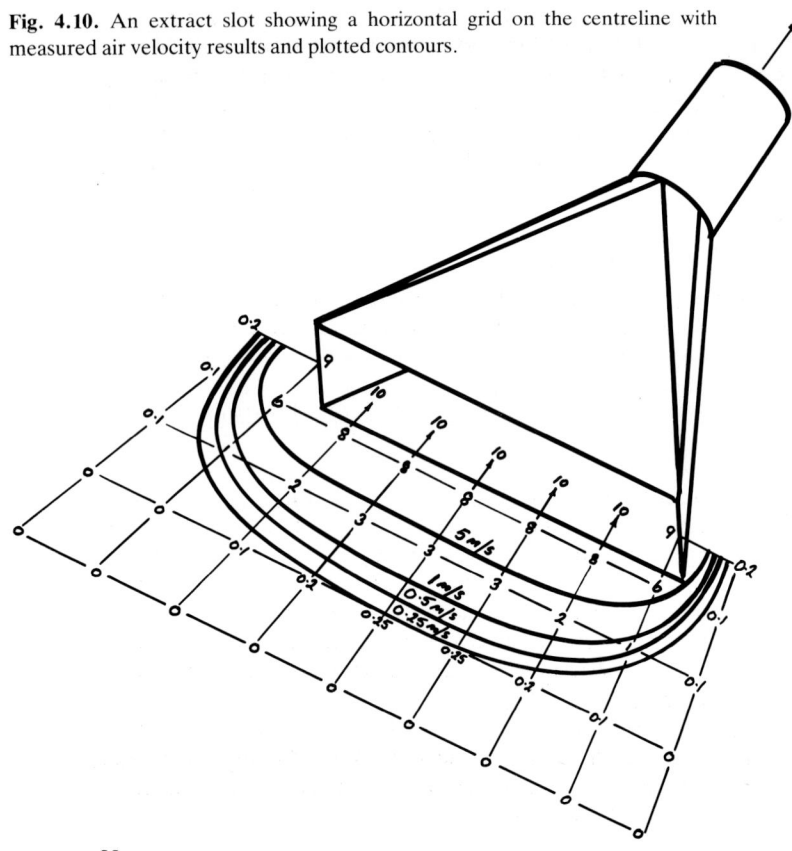

Fig. 4.10. An extract slot showing a horizontal grid on the centreline with measured air velocity results and plotted contours.

air stream and, if turbulence is caused by cross draughts due to the presence of external influences such as open windows, doors or the movement of people or vehicles, then much pollutant will escape the zone of influence of the suction inlet and be released into the general air.

The measurement of natural air infiltration rate in a room

Purpose

It is normally possible to obtain an indication of the airflow rate in a room which is mechanically ventilated as the majority of the air flowing will pass through the fan or duct serving the room but with rooms that rely on infiltration of air by seepage through the building fabric and around doors and windows it is much more difficult. It may be important to know the flow rate in order to calculate heat losses or to establish whether a room requires to be mechanically ventilated for a particular purpose or it may be useful to know just how 'leaky' the room is. The impracticability of trying to measure flow rates around doors and windows and cracks in the building fabric is obvious thus an alternative technique must be used. With most rooms the natural infiltration rate will depend upon external influences particularly the wind, therefore any result obtained must be related to wind speed and direction. An excercise such as this must be repeated several times over a variety of weather conditions in order to obtain a fair assessment of the natural flow rates as one single measurement is virtually meaningless.

Equipment required

A supply of tracer gas such as nitrous oxide, sulphur hexafluoride or radioactive Krypton K^{85}, a detector capable of measuring concentrations of the chosen gas at low levels, a desk fan, a timer, a sheet of log-linear graph paper and an anemometer.

Method

The method described here will be using the tracer gas K^{85} and a Geiger-Muller tube and counter as the detector but the procedure is essentially the same for whatever tracer gas and detector is used. The principle involved is to release a quantity of the tracer gas into the room and to measure how quickly it is diluted by the infiltration air.

1   Adjust the room to the conditions for which the measurements are needed, that is, shut the doors and windows or leave certain of them open as required.
2   Place the desk fan and Geiger-Muller tube on a table in the centre of the room and pass the cables out under the door. Connect the Geiger cable to the counter situated outside the room in the corridor or adjacent room and measure the background count.

Chapter 4

3 Start the desk fan which should be switched from outside the room.
4 Enter the room and break a 2mC ampoule of K^{85} on a tray provided to capture the broken glass and leave the room immediately.
5 Wait for five minutes to allow the desk fan and diffusion forces to thoroughly mix the gas with the air in the room.
6 Switch off the fan and measure the number of counts per minute from the Geiger counter. Continue to note the counts per minute at five minute intervals until the concentration of the gas has been diluted to a level approaching the measured background or for 45 minutes whichever is the shorter. (In the case of a tracer gas other than K^{85} being used note the concentration as measured by the detector every five minutes over a 45 minute period).
7 Re-enter the room and open all windows to clear the remaining gas.
8 Measure the wind speed and direction at a place that represents the true effect on the building such as the roof or at an open space adjacent.

Calculations

The airflow rate in number of air changes per hour is calculated from the expression:

$$N = \frac{1}{t}[\log_e C_o - \log_e C_t]$$

where: N = the number of air changes per hour,
t = the time between two chosen concentrations in hours,
C_o = the first concentration,
C_t = the concentration after time t.

Subtract the background count from all readings and choose two readings, one from the start of the test and one near the end and substitute in the above expression using t as the time interval between the two readings. Note that most scientific electronic calculators have a $\log_e$ or ln function which will simplify this calculation.

Example

Background count = 40,
Readings taken:

Time	Count	Corrected count
0 min	1520 per min	1480 per min
+5	861	821
+10	300	260
+15	256	216
+20	170	130
+25	105	65
+30	75	35
+35	65	25
+40	56	16

Taking counts at time = 0 and time = +35 min, $C_o = 1480$, $C_t = 25$, $t = 35/60$ hours.

Air change rate $N = 60/35\,[\log_e 1480 - \log_e 25] = 7$ air changes per hour.

An alternative graphical solution can be made by plotting on log-linear graph paper, the corrected count (or gas concentration) on the log scale against time on the linear scale and drawing a straight line through the points. The slope of this curve represents the number of air changes per hour.

To convert air changes per hour to $m^3\,s^{-1}$ multiply by the volume of the room in cubic metres and divide by 3600.

Calibration of an anemometer in an open jet wind tunnel

Purpose

To check the accuracy of an air flow measuring device and to provide a calibration chart for the range of the instrument.

Equipment required

An open jet wind tunnel as shown in fig. 4.11, orifice plates, manometer (inclined guage), dry bulb thermometer, anemometer stand as supplied with the tunnel or a retort stand, a barometer.

Fig. 4.11. Open jet wind tunnel (Airflow Developments Ltd).

Chapter 4   Method

1   Read and record tunnel air temperature in °C and room barometric pressure in mm of mercury.
2   Mount the anemometer to be tested 150mm from the discharge nozzle. Make sure that the direction of flow through the instrument is correct and that the face of the instrument is at right angles to the air jet and placed centrally. A template should be made to ensure this.
3   Select and mount an orifice plate for the velocity range to be tested.
4   Set up and zero the manometer and connect it by means of flexible plastic tubing to the pressure tappings on each side of the orifice plate, the high pressure side being the nearest to the fan.
5   Switch on the fan and by means of the controlling reostat adjust the air flow rate so that the anemometer dial reads a whole number.
6   Record the anemometer indicated flow rate and the orifice plate pressure difference as indicated on the manometer. Multiply this pressure reading by the appropriate scale factor for the angle of inclination of the guage.
7   Alter the airflow rate and repeat until four results are obtained from the orifice plate.

Table 4.2. Results sheet for open jet wind tunnel calibration test

Orifice plate	Diameter mm	Calibration constant A*	Recommended vel. range m s^{-1}	Indicated velocity V_i m s^{-1}	Orifice plate pressure diff. Δp Pa	True air velocity v_a m s^{-1}	Velocity correction V_c m s^{-1}
1	32.2	0.0325	0.2–0.9	1	1		
				2	2		
				3	3		
				4	4		
2	72.2	0.174	0.75–5.0	1	1		
				2	2		
				3	3		
				4	4		
3	101.6	0.356	2.5–10.0	1	1		
				2	2		
				3	3		
				4	4		
4	163.8	1.168	5.0–30.0	1	1		
				2	2		
				3	3		
				4	4		

*This constant is specific to an individual wind tunnel and orifice plate and should be supplied by the manufacturer or obtained by calibration.

8 Change the orifice plate and repeat so that four results are obtained from each orifice plate.
9 Record all results taken on a table similar to that shown in table 4.2.

Results and calculations

Calculate air density correction factor (d) from,

$$d = \sqrt{\left(\frac{760}{b} \times \frac{273+t}{293}\right)}$$

where: t = tunnel air temperature in °C, b = barometric pressure in mmHg.
Calculate the 'true air velocity (V_a) for each pressure reading from:

$$V_a = Ad\sqrt{\Delta p} \text{ m s}^{-1}$$

where: Δp = the orifice plate differential pressure in Pa
A = calibration constant for each orifice plate
Calculate the corrected velocity (V_c) for each reading from: $V_c = V_a - V_i$
Plot V_c against V_i on a graph whose ordinate are shown in fig. 4.12.

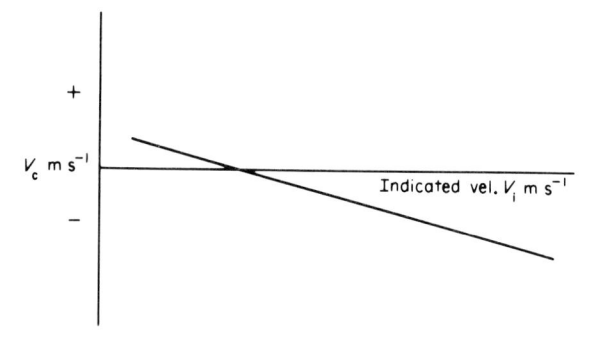

Fig. 4.12. Typical calibration chart for an anemometer.

Further reading

Fan Testing and Performance Prediction. In Daly BB, *Woods Practical Guide to Fan Engineering*. Woods of Colchester, 1979.
Testing of ventilation Systems. In *Industrial Ventilation*, American Conference of Government Industrial Hygienists, Lancing, Michigan.
BS 848, *Fans for General Purpose*, Part 1, Methods of testing performance. British Standards Institution, 1980.
Gill FS, Ventilation. In Waldron HA & Harrington JM (eds), *Occupational Hygiene*. Blackwell Scientific Publications, 1980.

CHAPTER 5
NOISE

Introduction

In modern environments it is difficult to find a situation where noise does not occur. Even people in offices and computer rooms experience noise levels which may cause concern and many industrial situations are noted for their continuous sound output.

The production and transmission of sound is complex and its understanding involves a good knowledge of physics and mathematics. However, recent developments in instrumentation are such that people whose knowledge of acoustics is limited can now provide for themselves a good assessment of the acoustical environment in which people work. Expert advice need now only be required if noise levels require reduction or a process is to be designed with minimum noise output.

In order to assess the health hazard of a noisy working environment it is necessary to measure the worker's exposure 'dose' of noise. This means assessing the sound intensity, the duration of exposure and the pitch of the sounds produced in the workplace.

Sound is produced at a range of pitches from a very low rumble or hum to a very high pitched squeal or hiss. The pitch is termed 'frequency' and is expressed in Hertz (Hz). This is the number of vibrations or pressure waves per second. The lowest pitch sound that can be heard by the human ear is at about 20Hz and the highest up to 18 000Hz (18kHz) for a young person. Each time the frequency is doubled the note will rise one octave. Middle C on the piano is 256Hz and C an octave above that is 512Hz. The ear is most sensitive to the frequencies of between 500 and 4kHz of which 500–2kHz is the frequency of speech. Unless a sound is a pure tone, which is unusual, most noises are made up of sounds of many frequencies and intensities and when assessing the intensity it may be necessary to discover what they are over the whole range of frequencies, that is, to measure the sound spectrum. For convenience it is usual to divide the sounds into octave bands and use a measuring instrument which assesses the intensities of all notes between the octaves and express it as a mid-octave intensity. The mid-octave frequencies chosen for this analysis are: 62.5Hz, 125Hz, 250Hz, 500Hz, 1kHz, 2kHz, 4kHz, 8kHz, and sometimes 16kHz. Thus a spectrum of a noise will quote the intensities at each of these mid-octave band frequencies. A technique for measuring a noise spectrum is given later.

Sound is pressure changes in the air which are picked up by the ear drum and transmitted to the brain. The pressure is normally measured in

Newton per square metre or Pascal (Pa). The quietest sound that can be heard is at about 0.00002Pa but at 25m from a jet aircraft taking off it is 200Pa which is 10 000 000 times greater. The intensity of sound is expressed in a unit known as the 'decibel' (dB) which is a convenient way of expressing a value which can have an extremely wide range.

The decibel compares the sound being measured with the threshold of hearing. Its value is obtained by dividing the pressure involved to produce one into the other and is expressed as a log scale (to base 10) because of the numerically large values obtained. The decibel is one tenth of a bel and is mathematically defined as:

$$\text{decibel dB} = 20 \log_{10} \left(\frac{p_a}{p_r}\right)$$

where p_a = the sound pressure of the sound being considered, and
p_r = the reference sound pressure at the threshold of hearing, that is, 0.00002Pa.

To illustrate this table 5.1 indicates typical everyday sound levels.

Table 5.1. Typical sound intensities

	Pressure Pa	Bels	Decibels
Threshold of hearing	0.00002	0	0
Quiet office	0.002	4	40
Ringing alarm clock at 1m	0.2	8	80
Ship's engine room	20	12	120
Turbo-jet engine	2000	16	160

Because the measurement of decibels is on a log scale, one sound is approximately twice as loud as the other when the difference between the two is 3dB. If two sounds are being emitted at the same time their total combined intensity is not the numerical sum of each separate intensity but must be added according to table 5.2.

Table 5.2. Simplified method for adding dB levels

Difference in dB	Add to the higher
0 or 1	3
2 or 3	2
4–9	1
10+	0

Chapter 5

Example

Two sounds at 90dB each will combine to have an intensity of 93dB, one at 90 and one at 92 will combine to an intensity of 94, one at 90 and one at 100 will combine to be 100, that is if two sounds are ten or more decibels different, then the lower intensity will not be added.

Weighting

As noise is a combination of sounds at various frequencies and intensities, the noise intensity can be either expressed as a spectrum, mentioned previously, or as a combination of all frequencies summed together in one value. As the human ear is more sensitive to certain frequencies than others, it is possible to make allowances for that in the electronic circuitry of a sound level meter. That is, certain frequencies are suppressed as others are boosted in order to approximate to the response of the ear.

This technique is known as weighting and there are A, B, C and D weightings available for various purposes. The one that is most usually quoted is the A weighting and instruments measuring sound intensity with that weighting give readings in dB(A). The weighting given to the mid-octave band frequencies for the dB(A) scale is given in table 5.3.

Table 5.3. Mid-octave band frequency corrections for dB(A) weighting

Frequency Hz	62.5	125	250	500	1000	2000	4000	8000
Correction dB	−26	−16	−9	−3	0	+1	+1	−1

The other aspect of noise dose which is to be measured is the duration of exposure of the worker to the noise at a particular intensity. This is simple if the noise produced is constant throughout the period of exposure then it is a matter of measuring the level and noting the time when the worker entered the area and when he left. However, most workplace noise fluctuates as different operations take place and various items of noise producing equipment or machinery are turned on or off. Also impacts such as hammering, rivetting and pressing cause short duration peaks of high intensity. Thus it is not easy to establish how long a worker has been exposed to each noise to work out a total dose. To do so it is necessary to note the duration of each noise level over the exposure period to build up a picture of the levels exceeded for a given percentage of the time. This is known as L_{10} for 10% of the time and L_{90} for 90% and so on, referred to here as L_x. Another and more widely adopted measure for workroom exposure is the 'equivalent continuous sound level' or L_{eq} which is the continuous noise level which would give the same total amount of sound energy as the fluctuating noise.

Until recently these measurements have been achieved by making very accurate tape recordings of the noise and playing them back through a statistical distribution analyser, but now instruments are available which will make the analysis electronically and provide a read-out at any moment of levels or noise dose that have accumulated from the instant that the meter was turned on.

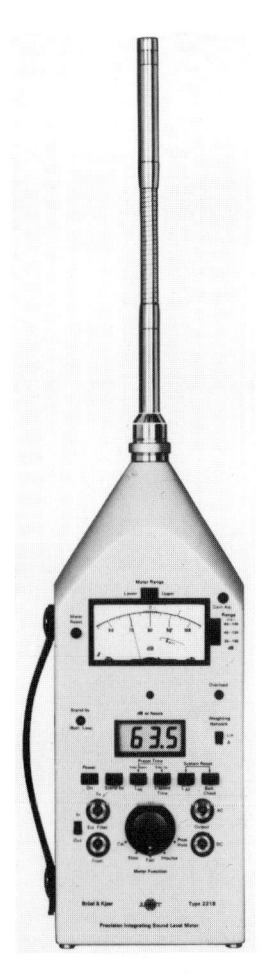

Fig. 5.1. Precision grade integrating sound level meter (Bruel and Kjaer (UK) Ltd).

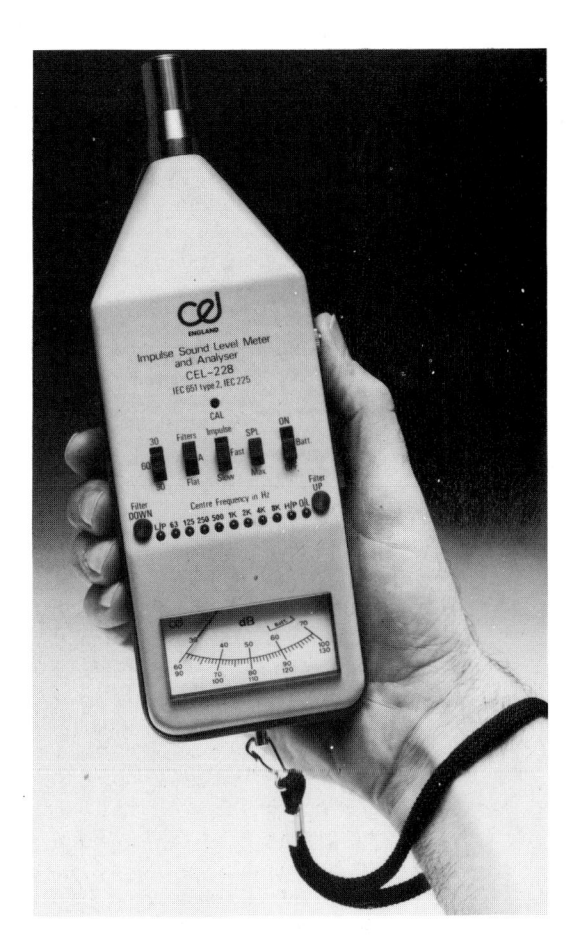

Fig. 5.2. Impulse sound level meter with integral octave band analyser (Computer Engineering Ltd).

Chapter 5

Equipment available

Sound level meters

Essentially a sound level meter consists of: a microphone which converts the sound waves to electrical impulses; some electronic circuitry which modifies and amplifies the signal according to the demands of the user, and an output section to display the signal in units of sound as required. Portable sound level meters are battery powered, thus the battery charge condition must also be displayed. Until recently the quality of portable field meters varied from a simple 'industrial' grade meter giving a level in dB(A) to a 'precision' grade having many facilities, although the grading was based on the degree of precision provided. New international standards which have also been adopted by the British Standards Institution have specified four grades or 'types' of precision. Type 0 is the most precise 'laboratory' grade instrument having the most stringent tolerances of performance whilst type 3 has the widest tolerance. The quality and facilities which a sound level meter offers is reflected in its cost. Prices start at around £100 for the simplest instruments and rise to several thousand pounds for the most precise ones having a wide range of facilities.

Recent work has shown that impulse noise and short term transient peaks of sound have an important effect upon noise exposure. Most traditional sound level meters do not have sufficient dynamic range or a fast enough response time to correctly record these short term events. International standards now specify requirements for impulse precision sound level meters which respond very quickly. The most recently developed meters comply with the new standards but older designs which are still on the market do not. Care should be taken when purchasing new instruments to ensure that the type is suitable for the purpose required.

Weight and size also vary, the ones having the most facilities being the heaviest and most bulky. Where important decisions are to be made based on noise measurements or if legal arguments are to be based on results, then the precision grade instrument must be used, preferably handled by an experienced qualified acoustician.

Noise dose meters

The sheer size and weight of a meter which has a noise dose facility precludes its use for personal measurement. Therefore, small noise dose meters are available which can be easily carried by a worker throughout the shift and which will not hinder the task to be performed. These will sense all noise levels to which it is exposed making a record against time and storing the information for retrieval on demand. Dosemeters can either have an integral microphone, the whole unit designed to be worn

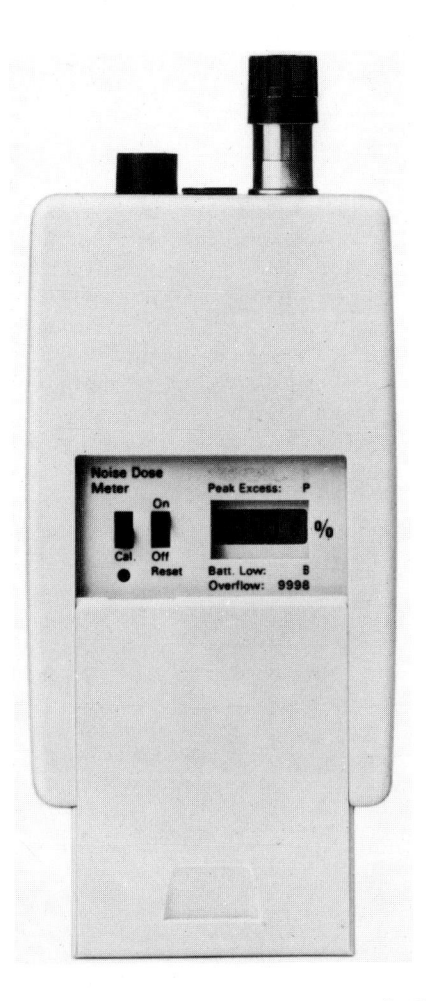

Fig. 5.3. Personal (ISO) noise dose meter (Bruel and Kjaer (UK) Ltd).

in the breast pocket, or they can have a separate microphone which can be clipped to the lapel with a lead carrying the signal to a small box containing the battery and electronics which can be carried in a pocket or hung on a belt. An example of a noise dosemeter is shown in fig. 5.3.

As with sound level meters, noise dose meters can be either of precision or industrial grade.

Calibration

In order to check that the meter is reading correctly it is necessary to have a source of sound of a known intensity and frequency which can be

Chapter 5

introduced through the microphone excluding all other sounds except the calibration tone. All meters have an adjusting screw which can be turned to make the display read the intensity of the calibrator.

Most calibrators for portable meters are battery powered producing the calibration tone electronically. They must be of a tight fit around the microphone of the meter to ensure no other sounds are being received at the same time. To this end rubber O-rings are used as seals and some have adapters to fit various microphone sizes. Calibration tones vary in intensity from one manufacturer to another although 94dB at 1000Hz is the most usual. If dB(A) scales are to be checked then they must be at a frequency of 1000Hz. There is a calibrator available known as a 'pistonphone' which produces a tone of 124dB at 250Hz and this is suitable for calibrating an octave band analyser at that frequency.

It is important to check the meter regularly with the calibrator and at least once before starting a series of readings or at the start of a working shift. It is also important to ensure that the calibrator battery is in good order and each manufacturer gives instructions as to how to achieve this.

To measure a steady workroom noise

Aim

Workroom noise produced from machines or items of equipment can be a source of irritation, stress and/or can lead to hearing damage. It is important to be able to measure noise levels at various workplaces to establish whether it is necessary to take remedial action to control the workers' exposure and to identify the main sources of noise. Following on from that it may be necessary to designate certain areas of the factory as hearing conservation areas in which it is unwise to enter without wearing suitable protection.

Equipment required

A sound level meter having a dB(A) weighting capability, a calibration device suitable for the microphone on the meter, that is, of the same diameter to ensure a good fit, a small screwdriver, a tripod capable of holding the meter may also be useful but not essential.

Method

1 Check the battery output by switching on the 'battery test' switch, the needle should swing to beyond a designated mark on the meter scale. The position of the test switch varies for different instruments but is usually associated with the power on/off switch and is always clearly marked. If the needle fails to reach the mark the batteries should be replaced. It is wise to test the battery before leaving for the survey unless good spares are carried.

2   Switch on the instrument and allow it to warm up for at least two minutes.

3   Calibrate the instrument as follows: remove the microphone cover, fit the calibrator over the microphone and set the scale to dB(A) and to the correct range* for the output of the calibrator. If the instrument has a 'fast' and 'slow' response switch set it to 'fast'. Turn on the calibrator and observe the reading on the meter. If it does not read exactly the calibration value adjust the needle by turning the calibration screw using the small screwdriver. The adjusting screw will be labelled in a variety of ways, e.g. 'Adj', 'Gain Adj' or 'CAL'.

4   To measure the noise exposure remove the microphone cap, turn on, switch to fast response and hold the instrument at arms length away from the body keeping it at least one metre above the floor. If it is a worker's exposure that is to be measured hold the microphone close to each ear of the worker but pointing towards the source and note the reading on each side. If the meter is fluctuating too much to obtain a readable value then switch it to 'slow' response and read again. If the needle rises above or falls below the range* of the meter, switch the range control to the appropriate value. In order to minimise the sound blocking effect of the measurer's body, if a tripod is available mount the instrument on it and stand at least 0.5m away from it when reading the meter. Some instruments even have their microphone on a length of cable so that it can be mounted away from the meter for this very purpose.

5   If the worker is at a noisy machine it may be useful to obtain a background noise, therefore repeat 4. above with the machine switched off. If there is little difference in the noise level between the background only and the machine plus the background (less than 3dB) then other sources of noise may be just as important.

Noise

Results

Official levels of continuous noise, or its equivalent energy, that are set to avoid hearing damage vary from one country to another. At present (1982) in Great Britain the standard is 90dB(A) for 8 hours per day but some countries set the more stringent standard of 85dB(A). Therefore it would be wise to take precautions at any value above 85dB(A). Annoyance and stress which may lead to accidents or other problems are produced at much lower levels therefore complacency should not necessarily be felt if lower values than the current standard are obtained.

*Meters vary in the way they indicate the reading, some show a value on a dial to add to or subtract from the setting of a range switch, some indicate directly on a dial according to the range to which it is switched and some give a digital read out of actual dB.

Chapter 5

The measurement of the spectrum of a continuous noise by octave band analysis

Aim

In order to ascertain the range of frequencies represented in a problematical workroom noise it is necessary to break that noise down into bands of pitch and to measure the intensity of each band. In this way it is possible to establish what frequencies, if any, in the spectrum are the loudest so that the effect of the noise can be assessed. For example, if the loudest sounds are in the speech frequencies (500Hz–2kHz) it will be clear that speech communication will be difficult or impossible and if the loudest sounds are at the higher frequencies of 1–6kHz then those people exposed to it are more likely to suffer temporary or permanent hearing damage than if the lower pitches are represented. From this information the acoustic engineer can decide what remedial action needs to be taken as different spectra require different attenuation techniques to reduce the levels. It is generally easier to suppress high frequency sounds than low rumbles or hums which, although are less damaging to hearing, are nevertheless extremely irritating particularly if they continue for long periods of time. Also from the spectrum, if hearing defenders are to be prescribed then the type most suited to the noise characteristic can be specified.

Equipment required

A precision sound level meter with an octave band filter set incorporated or added (a third octave band instrument could be used to provide more information but is more expensive and less common); additional filter sets are usually screwed on to the bottom of the meter*; a calibration device suitable for the microphone on the meter, that is, of the same diameter to ensure a good fit; a small screwdriver; a tripod capable of holding the meter may also be useful but is not essential.

Method

1   Check the battery output by switching on the 'battery test' switch, the needle should swing to beyond a designated mark on the meter scale. The position of the test switch varies for different instruments but is usually associated with power on/off switch and is always clearly marked. If the needle fails to reach the mark the batteries should be replaced. It is wise to test the battery before leaving for the survey unless good spares are carried.

2   Switch on the instrument and allow it to warm up for at least two minutes.

*With the B & K 1613 filter set ensure that the weighting switch on that piece of equipment is in the 'off' position.

3 Calibrate the instrument as follows: remove the microphone cover, fit the calibrator over the microphone and set the instrument to 'ext filter' or 'filter' and set the response to 'fast'. Select the meter range* to the correct one for the output of the calibrator. On the filter set, switch the octave band selector to the frequency of the calibrator. Turn on the calibrator and observe the reading on the meter. If it does not read exactly the calibration value adjust the needle or digital display by turning the calibration screw using the small screwdriver. The adjuster will be labelled in a variety of ways, e.g. 'Adj', 'Gain Adj' or 'CAL'.

4 Locate the position where the measurements are required to be taken, this may be at the place where the worker is stationed or if the point of maximum noise level is required then it will be necessary to ascertain this by using the dB(A) scale as described previously and measuring in various places in the workroom until this point is established.

5 To obtain the sound spectrum: remove the microphone cover, turn on the instrument, switch to 'Ext filter' or 'Filter' and to 'slow' response and hold the instrument at arms length away from the body pointing towards the noise source. Turn the external filter band selection knob to 31.5Hz and note the reading. If the scale on the meter is at the wrong range select the correct one.* Repeat for each filter band setting up to 16kHz. In order to minimise the sound shielding effect of the observer's body the instrument can be mounted on a tripod and read from a distance of at least 0.5m.

6 As a check, turn off the external filter and set the instrument to dB(A) and measure the A weighting level.

Results

Plot the results obtained from the measurement on a copy of the Noise Rating (NR) curves given in fig. 5.4. Join up the points with straight lines (not a smooth curve). Examine the shape of the spectrum to establish whether there is a steady output at all frequencies or whether certain frequencies predominate. It is beyond the scope of this text to provide ways of judging the results as the field of acoustics is very specialised. The reader is advised to seek further advice by perusing a good acoustics text or by consulting an acoustician with experience in the field of human response to noise. Nevertheless the NR curves are given to provide some guidance. The noise rating of the spectrum obtained can be said to be the same as the curve immediately above the highest point on the spectrum drawn—see the example in fig. 5.4.

The dB(A) of the spectrum can be obtained by adding the intensity of

---

*Meters vary in the way they indicate the reading. Some show a value on a dial to add to or subtract from the setting of a range switch, some indicate directly on a dial according to the range to which it is switched and some give a digital read out of actual dB.

Chapter 5

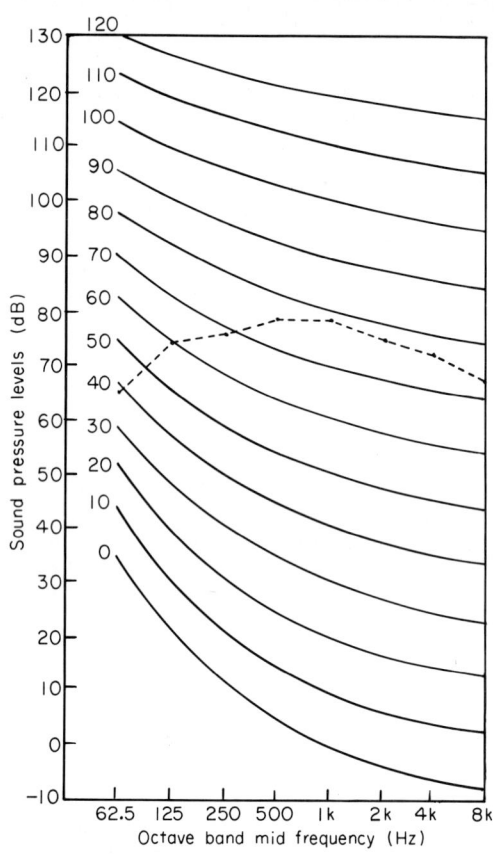

Fig. 5.4. Noise rating curves showing the plotted spectrum from the example.

adjacent bands until a single value is obtained. Addition of sound levels is quickly but approximately done using table 5.2. Draw up a record sheet as shown in the example in table 5.4 and write the results on the second line.

Table 5.4.

Octave band mid-frequency Hz	63	125	250	500	1000	2000	4000	8000
Measured levels dB	65	74	76	78	78	75	72	67
Correction for dB(A)	−26	−16	−9	−3	0	+1	+1	−1
Result	39	58	67	75	78	76	73	66

Addition from table 5.2

58   76   80   74
  76   81
     82dB(A)

104

Example

With several machines running in a duplicating room in an office building a sound spectrum was obtained as shown in the second line of table 5.4. The measured dB(A) level was 82. The sound spectrum is plotted on the NR curves shown as a broken line.

The calculated dB(A) agrees with the measured value. It can be seen that the plotted spectrum just lies below the NR78 curve, the inference being that this is a very high noise level to be experienced in an office situation and is above what might be desirable even in an industrial workplace.

Recommended noise ratings

Broadcasting or recording studio	NR15
Concert hall or a 500-seat theatre	NR20
Class room, music room, TV studio, large conference room, bedroom	NR25
Small conference room, library, church, cinema, courtroom	NR30
Private office	NR40
Restaurant	NR45
General office with typewriters	NR50
Workshop	NR65

To measure the L_{eq} of a fluctuating workroom noise

Aim

The meaning of L_{eq} is explained on page 96 but essentially it is the equivalent steady level in dB(A) of a fluctuating noise over a period of time. It is used to establish the risk to which a worker is exposed when subjected to a continually changing noise over the course of a working period. From the result it will be possible to decide whether or not action is required to be taken to reduce the exposure either by: reducing the noise intensity, reducing the duration of exposure or providing some form of hearing protection.

Equipment required

An integrating sound level meter, a calibrating device suitable for the microphone on the meter, that is, of the same diameter to ensure a good fit, a small screwdriver, and a tripod capable of holding the meter.

Method

1 Check the battery output by switching on the 'battery test' switch, the needle should swing to beyond a designated mark on the meter scale.

Chapter 5

The position of the test switch varies for different instruments but is usually associated with the power on/off switch and is always clearly marked. If the needle fails to reach the mark the batteries should be replaced. It is wise to test the battery before leaving for the survey unless good spares are carried.

2 Switch on the instrument and allow it to warm up for at least two minutes.

3 Calibrate the instrument as follows. Remove the microphone cover, fit the calibrator over the microphone and set the instrument to dB(A) and to SEL or SPL. Select the meter range to that appropriate for the calibrator. Turn the calibrator on and observe the reading on the meter. If it does not read exactly the calibration value adjust the needle or digital display by turning the calibration screw using the small screwdriver. The adjuster will be labelled in a variety of ways, e.g. 'Adj', 'Gain Adj' or 'CAL'. On some meters it is possible to switch to L_{eq} mode to check that the reading of the meter is the same but when the calibration device stops emitting the reading should then drop to below the calibration level.

4 Locate the instrument in the position where the measurements are required to be taken which would normally be at a work station or close to a worker. Because it is to be left for a period of time it is necessary to mount it on a tripod or on some vibration free surface. After ensuring that the meter is turned on, the microphone cover removed and the weighting switch at 'A', switch to L_{eq} at the moment when the worker's exposure to the noise starts.

5 At the end of the period note the value of the L_{eq} as displayed. Some instruments show this continuously throughout the test period but others require a display button to be pressed to provide a digital readout at any time during the test. One instrument displays the duration of the test period in hours when the 'elapsed time' button is pushed.

Results

Standards of continuous noise or its equivalent energy, L_{eq}, that are set to reduce hearing damage vary from one country to another. At present (1982) in Great Britain the standard is 90dB(A) for an eight hour period of exposure. A scale of reduced exposure times for levels above this value are shown in table 5.5. In order to minimise the risk of workers suffering hearing damage it is important to ensure that this table is adhered to or some form of suitable hearing protection provided.

The use of a personal noise dosemeter

In order to assess the degree or auditory hazard to which a worker is exposed during the course of his or her job it is useful to monitor the worker rather than the workplace. This particularly so if the worker

Table 5.5

Limiting dB(A)	Maximum duration of exposure
90	8 hours
93	4 hours
96	2 hours
99	1 hour
102	30 minutes
105	15 minutes
108	7 minutes
111	4 minutes
114	2 minutes
117	1 minute
120	30 seconds

moves about into various levels of noise exposure or where the use of a bulky integrating sound level meter is difficult or impossible due to the nature of the operations taking place. Personal dosemeters are designed to be socially acceptable and cause minimal interference when worn as they are small and light in weight.

Equipment required

A noise dosemeter, a calibration device suitable for the dosemeter, a small screwdriver.

Method

Dose meters from different manufacturers are so widely different that it is not possible to be general in describing their use. Therefore the reader is advised to follow closely the instructions provided with the instrument. However, some general comments are provided here for guidance of users.

1 The read-out displays of dosemeters vary, the most recent types give a digital display using liquid crystal diodes but earlier models used a binary code display involving sixteen lights arranged in groups of four, each group representing: units, tens, hundreds and thousands. The instruction books give guidance on interpretation of the display.

The result of the measurement is expressed as a percentage of the allowed daily L_{eq} exposure, thus an exposure of 90dB(A) for eight hours would be 100% as would 93dB(A) measured for four hours or 96 for two hours. Formulae and conversion charts are given in the handbooks to convert the recorded percentage to L_{eq} knowing the duration of the test. Also some meters display a peak level warning which indicates when a specified dB(A) value has been exceeded during the test.

2 Switching of the meter is arranged to provide unwanted interference and to make the device as tamper proof as possible. This is achieved in a variety of ways. In one model the instrument is turned on or off and the display actuated by depressing microswitches set below the level of the cover by poking a narrow allen key through holes in the casing. In another model these operations are actuated by magnetically operated reed switches, a magnet being provided for the purpose whilst a third type has all its switches and displays hidden by a sliding cover which can be sealed by an adhesive label which indicates whether the seal has been disturbed.

3 As a dosemeter is a device which stores information over a period of time it is important to ensure that all previous measurements are cleared from the memory, thus a resetting procedure must be adopted if the read-out is not showing zero.

4 The instrument must be calibrated before use as directed by the manufacturer.

5 The microphone should be placed as close to the ear as possible, either by clipping to the lapel or by attaching it to the rim of a helmet. It should be noted that one ear of the worker may be exposed to a louder noise than the other and in this event the microphone should be attached on that side of the worker's body. Care should be taken to ensure that the microphone is not knocked in any way in the course of the worker's actions as artificially high readings may result. Microphones should also be protected by a dust cover when in use although it is necessary to remove it when calibrating.

Results

After converting the displayed percentage to a true worker exposure a judgement can be made as to the likely harmful effects using table 5.5. If necessary action should be taken to minimise worker exposure to the noise.

Further reading

King IJ. Noise. In Waldron HA & Harrington JM (eds), *Occupational Hygiene*. Blackwell Scientific Publications, 1980.
Burns W. *Noise and Man*. Murray, 1973.
Taylor R. *Noise*. Pelican, 1975.
Department of Employment. *Code of Practice for Reducing the Exposure of Employed Persons to Noise*. HMSO, 1976.
Sharland I. *Practical Guide to Noise Control*. Woods of Colchester Ltd, 1979.
Webb DJ (ed). *Noise Control in Industry*. Sound Research Laboratories, 1976.
BS 5969. *Specification for Sound Level Meters*. British Standards Institution, 1981.

CHAPTER 6
LIGHT

Introduction

It is important that workplace lighting is maintained at a good standard, that is, lighting levels must be sufficiently high to enable workers to clearly see their tasks but not too high as to cause glare or dazzle. Poor workplace lighting not only creates eye strain, particularly if the task to be performed contains small detail, but can cause fatigue leading to errors in the work and an increased accident risk.

Sources of illumination can be divided into two: natural (daylight) and artificial (usually by means of electric lamps). Very few workplaces rely solely upon daylight whereas many are entirely artificially illuminated. Different levels of illumination are required for different tasks, thus workplace lighting must be designed for the type of work to be undertaken. Unfortunately work patterns change, sources of illumination deteriorate with age particularly in industrial situations, that is, windows and light fittings accumulate dirt which reduces the amount of light emitted and surfaces become dirty, reducing the amount of light reflected from them. This often occurs so gradually that it goes unnoticed. Therefore it is prudent for workplace lighting levels to be measured from time to time and the results checked against recommended standards.

However, it must be stressed that the presence of the correct level of illumination does not necessarily mean that the workplace is properly lit. The position of the source of light in relation to the worker and the workpiece may seriously affect the way the task is seen. The appearance of solid objects will be influenced by the direction from which the light comes and unshielded sources of light will cause glare if they appear within the field of view. Also if the work involves the identification of different coloured materials then certain sources of light can alter colours, sometimes making two different colours appear the same.

Units used in lighting

Luminous intensity. Symbol I, unit, candela, which is the power of a source to emit light, for example, a lamp may emit 100 candela.

Luminous flux. Symbol Φ, unit, lumen, which is the luminous flux emitted within unit solid angle (one steradian) by a point source having a uniform luminous intensity of one candela.

Illuminance. Symbol E, unit, lux (or lumen m^{-2}) which is the luminous flux density at a surface. This is the unit which is used to express the lighting levels in a workroom.

Luminance. Symbol L, unit, candela m^{-2}, which is the intensity of light emitted in a given direction by a unit area of luminous or reflecting surface.

Reflectance of a surface. Symbol R, unit, percent, which is the ratio of the luminance of a surface and the mean illuminance in the room expressed as a percentage.

Daylight factor. Symbol DF. Where daylight passing through windows is used to illuminate an indoor workplace, less than one tenth of the outside light is likely to reach that place and that amount decreases with distance from the window. It is useful to compare the inside illuminance due to daylight with that prevailing outside at the same time on a percentage basis. For this purpose a value known as the 'daylight factor' is used which is defined in Great Britain as: the percentage of daylight illumination at a given point on a plane in a building relative to that simultaneously prevailing outside under an unobstructed, uniformly overcast sky. It is expressed mathematically as:

$$DF = \frac{E_i}{E_o} \times 100 \text{ per cent}$$

where: E_i = the illuminance at a point inside
E_o = the simultaneous illumination outside.

The daylight factor at a point in a room is constant being dependent upon the solid angle subtended to the sky as viewed through the window or windows. Thus once the daylight factor of a point in a room is determined it is always possible to predict its illumination level knowing the outside illuminance at a particular time by rearranging the above expression:

$$E_i = \frac{DF \text{ per cent} \times E_o}{100}$$

In South East England the monthly average outside illumination at noon in July is 34 000 lux and at the same time in December is 7800 lux. The standard taken for the average exterior illuminance in this country is 5000 lux, thus when such a level occurs, a daylight factor of 1 per cent at a point inside a building would produce an average of 50 lux (1 per cent of 5000).

Diversity factor. This describes how even the distribution of light is in a room and is normally expressed as the minimum illuminance divided by the mean illuminance.

Utilisation factor. Symbol UF, unit, lumen/watt, which indicates how effective the sources of artificial light are in relation to the total amount of power provided by the lamps. It is calculated from:

$$UF = \frac{\text{mean room illuminance (artificial light only) in lux} \times \text{area of floor (m}^2\text{)}}{\text{total wattage of lamps}}$$

Maintenance factor. This indicates the condition of the lamps in relation to the amount of output the lamps produce based upon the utilisation factor. It is a value that will decrease as the lamps age or become dirty. It is calculated from the expression:

$$\text{maintenance factor} = \frac{\text{illuminance} \times \text{floor area}}{UF \times \text{total wattage of lamps}}$$

Equipment available

Photometers

These are photoelectric devices that consist of a photocell which converts light to an electric current. It is connected to a moving coil meter which indicates the current as lux. The most suitable type for workplace measurements should have a range of 0–2500 lux and should have the photocell separate from the meter but connected via a length of cable. This arrangement allows the meter to be read without the observer overshadowing the cell. The photocell should be corrected to take into account the effects of light falling upon it from an oblique angle (cosine corrected) and preferably colour corrected to allow measurements to be taken over a wide range of lamps and with daylight. Manufacturers of these meters do provide colour correction charts for use with instruments which approximate the illuminance. Typical colour correction factors are given in table 6.1. The accuracy of photometers should be checked from time to time using sources of known intensity. One such instrument is illustrated in fig. 6.1.

Table 6.1. Colour correction factors of a typical photometer

Light source	Multiply reading by
Fluorescent: warm white	1.27
daylight	1.20
colour matching	1.05
natural	1.14
Sodium low pressure	1.30
Sodium	1.19
Mercury vapour	1.26

Chapter 6   Hagner Universal Photometer

This is an instrument for measuring luminance or the brightness of light given off or reflected from a surface. It consists of a photoelectric cell coupled to a meter calibrated in candela m^{-2}. There is a viewfinder similar to a camera through which the surface to be measured is observed and the cell only responds to the light emitted from the surface contained within the field of the viewfinder. Thus to make a measurement the instrument has to be held up to the eye and pointed at the surface being tested as shown in fig. 6.2.

Daylight factor meter

These instruments normally consist of a selenium barrier-layer photocell cosine corrected by means of a filter which compensates for light reflected from the detecting cell surface above it and connected by way of sensitivity control to a micro-ammeter. A hinged louvred mask closes over the cell for outdoor use which reduces the overall illumination of

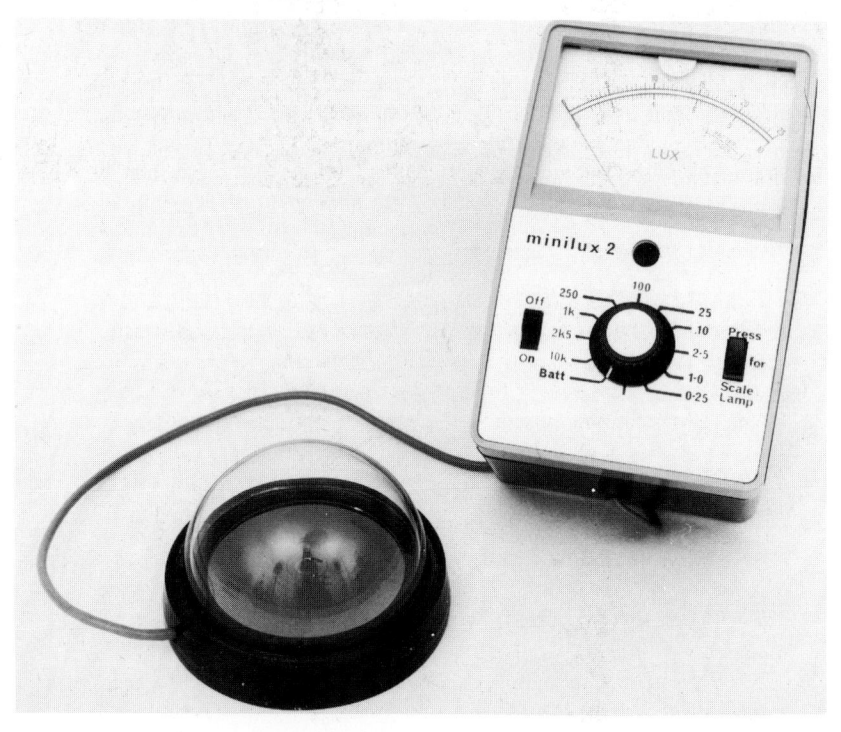

Fig. 6.1. Photoelectric photometer (Salford Electrical Instruments Ltd).

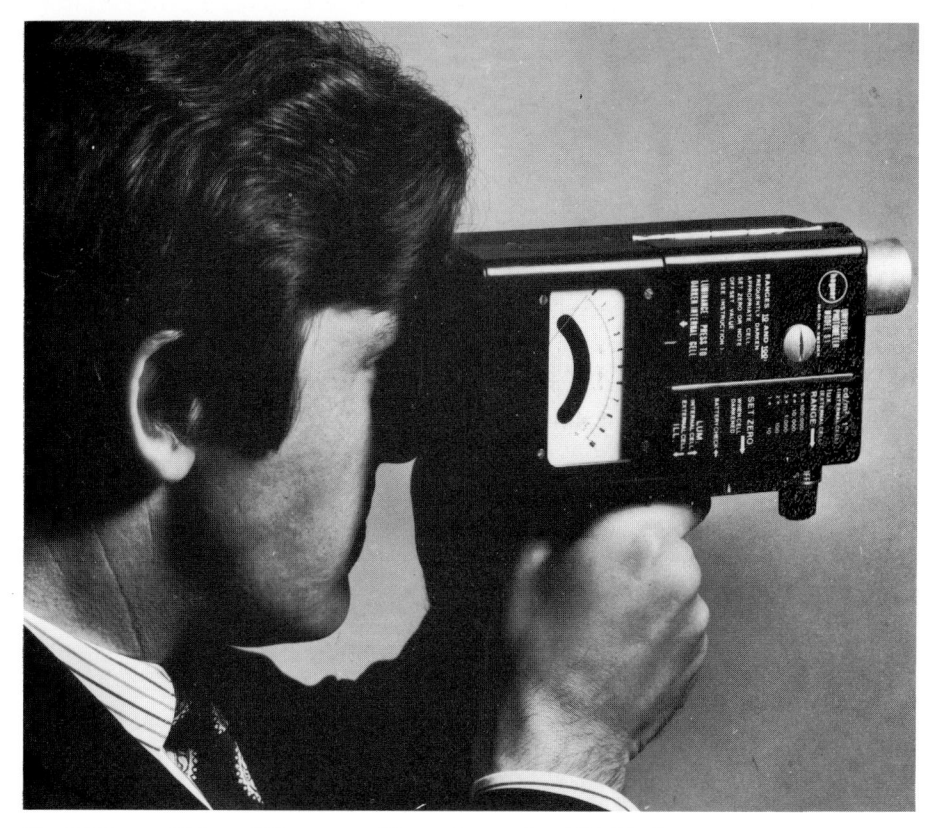

Fig. 6.2 Hagner Universal Photometer (Hagner International (UK) Ltd).

the photocell and admits light only from an elevation of 40°–50°, that is from that zone of the sky where the brightness is numerically equal to the illuminance of the whole sky. Indoor readings are taken with the mask hinged clear of the photocell for increased sensitivity with a non-directional characteristic. An instrument of this type is illustrated in fig. 6.3.

To measure the daylight factors in a room

Aim

Many workplaces rely upon daylight for the majority of the days' illumination, artificial light being provided only when the worker considers it is too dark to see properly. There is often a tendency for the person to delay turning on the lights until the illumination has reached such a low level that the work is affected or a higher than normal risk of

Chapter 6

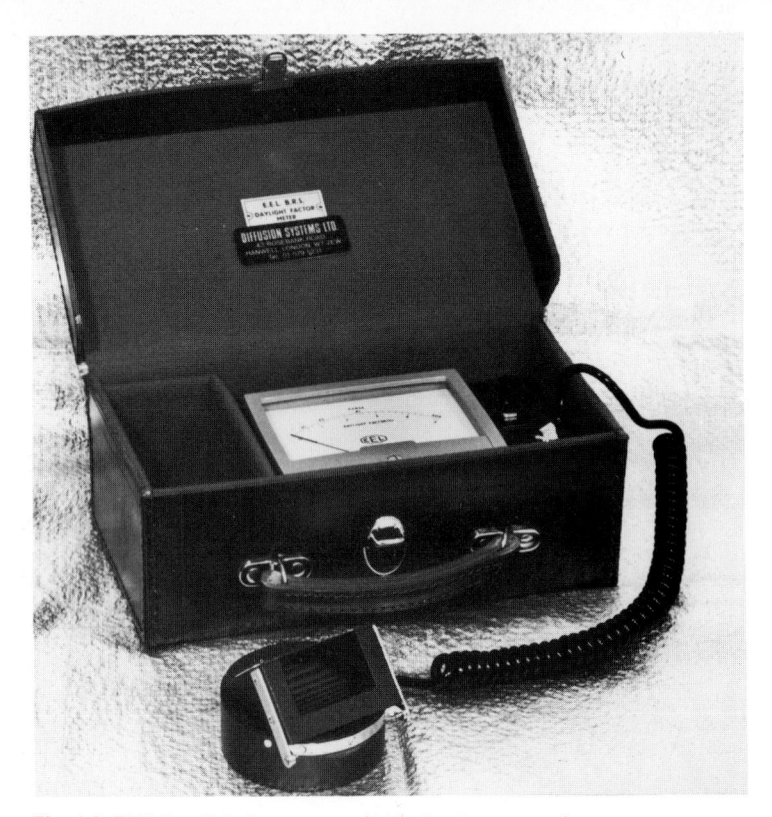

Fig. 6.3. EEL Daylight factor meter (Diffusion Systems Ltd).

an accident occurring is present. Therefore it is wise to provide some supplementary artificial lighting to boost the daylight entering through the windows which may be operating all of the time or switched on under the influence of a light sensor. Some scheme or programme of illumination control may have to be calculated based upon the measurement of daylight factors. Also the amount of daylight reaching a workplace may change due to the presence of external changes affecting the amount of light falling upon a window for such reasons as the erection of a new building or item of industrial plant. There may also be the need to check for excessive daylight causing glare on some workplace for which the remedy is shading from such items as venetian blinds but which then may affect the daylight illumination on some other part of the room.

For these and other reasons it is useful to be able to check the amount of natural daylight falling upon a workplace.

Equipment required

A daylight factor meter, some graph paper, a tape measure.

Method

Measure the room size and sketch a plan of it on the graph paper, noting the positions of workplaces and items of equipment. The windows should be measured for effective area and those values marked on the sketch at the appropriate positions.

1   Close the louvred mask over the photocell of the meter.
2   Stand outside or by an open window and hold the instrument at a convenient height to read the scale, direct the louvres towards an unobstructed area of overcast sky. Note that daylight factors can only be measured under an overcast sky.
3   Adjust the sensitivity control so that the meter scale reads X1 or X2, that is, line up the meter needle with the line indicating X1 or X2.
4   Return to the room to be measured and switch off all artificial lighting. Swing the louvred mask completely clear of the photocell (i.e. open) and taking care not to obstruct the daylight, hold the meter at various positions in the room and note the reading. It is advisable to move away from the window in increments of 0.5m, noting the results as laid out in the result sheet in table 6.2. The meter will read directly in terms of percentage daylight. It may be necessary to alter the sensitivity of the meter.
5   Repeat this for various traverses across the room so that the whole room area is covered. It is suggested that the traverses be 1m apart.

Table 6.2. Suggested results sheet for daylight factors

Distance from window wall m	Daylight factors in traverse									
	1	2	3	4	5	6	7	8	9	………
0.5
1.0
1.5
2.0
2.5
3.0

Results

Transfer the results to the sketch and draw contours through points of equal daylight as shown in fig. 6.4.

Chapter 6

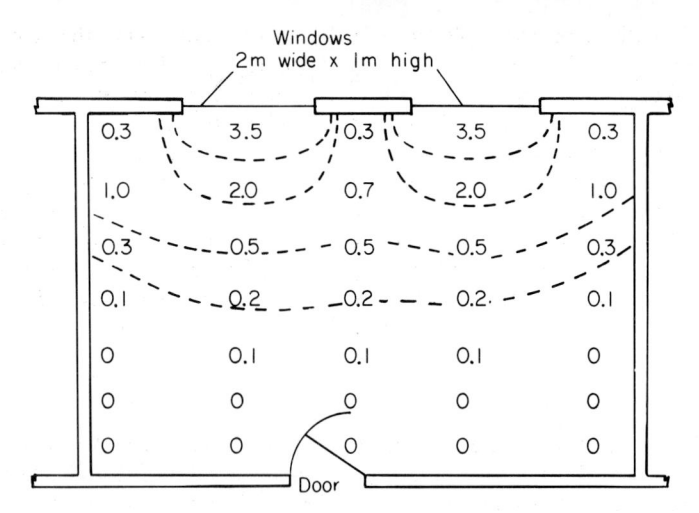

Fig. 6.4. Typical room plan showing daylight factors and contours.

To undertake a lighting survey of a workroom

Aim

In order to fully understand the distribution of light in a room and to determine whether workplace illumination levels are suitable and sufficient for the work to be undertaken, two stages of assessment are required. The first is to complete a subjective examination of the general illumination and the second is to follow that up with some organised measurement of lighting levels.

Equipment required

A portable photoelectric photometer, a visual assessment form, plain and graph paper, tape measure, pencils and, if possible, a camera.

Method

Preliminaries

1   Draw a sketch plan of the room to show principle working surfaces, windows, light fittings and other relevant features. This should be done to scale if possible. It may be necessary to draw separate floor and ceiling plans to avoid confusion between the furniture and the light fittings, these can be done on tracing paper or translucent sheets so that one can overlay the other. In order to assist the memory, section drawings can be made through windows to show their dimensions and the positions of

light fittings or an 'exploded' sketch can be made giving the same information. It is also useful to take a photograph of the room from various positions but failing that an isometric sketch can be made on which wall colours and the nature of surfaces can be noted.

2   Measure the major dimensions of the room and note them on the sketch plans. Note and record also the nature of the lighting in use, whether it is daylight or artificial or a mixture of both and record the type of lamp, its wattage and the type of diffuser or holder in use. The state of the room should be noted with regard to cleanliness, that is, observe whether the light fittings, lamps and windows are clean or dirty and the reflectances of the principle room surfaces.

Visual assessment

Having now spent some minutes in the room it is useful to make a visual assessment. Try to identify any specific aspects of the lighting installation which should be examined in detail. Decide if illuminance on working surfaces appear to be satisfactory and, if not, determine the reason why and in which parts of the room. Ascertain whether the present use of the space differs from the original purpose for which the lighting was designed and determine what the illuminance levels and the daylight factors should be for such a purpose built room from codes of practice such as the IES Code (see further reading).

Determine whether there are any undesirable shadows or reflections on the work. Notice whether the lighting fittings or windows cause discomfort or disability glare when seen separately or together and observe whether the windows are obstructed by internal furnishings or equipment and by outside trees, walls or other buildings which may affect the illumination.

If the nature of the work requires the recognition of different colours, note whether the colour rendering of the lights is satisfactory. This may have to be done by removing some of the coloured material to the window or outside to observe it under natural daylight to see whether the colour changes.

Notice whether there is any flicker from discharge lamps including fluorescent tubes and if any stroboscopic effects on moving machinery are present (rotating items may appear stationary if running at the same speed as the mains frequency).

The following check list may be helpful to undertake this subjective visual assessment:
1   Adequacy of lighting,
2   Suitability of lighting to function of room,
3   Shadows on work surfaces,
4   Reflections affecting work stations,
5   Glare from windows and lights:
    (a) causing discomfort,

Chapter 6

    (b) causing disability,
6 Obstructions to windows from inside or outside,
7 Colour rendering,
8 Cleanliness of light fittings, windows and wall surfaces,
9 Flicker,
10 Distribution of light over the whole room,
11 General impressions of whether satisfactory or not.

It is useful to chat to the people occupying the room to establish their subjective feelings about the lighting levels and whether any discomfort is experienced. They may also be in a position to suggest where improvements might be made.

Measurement

If the weather conditions are correct, that is, with an overcast sky, the measurement of illumination levels can be preceeded by a daylight factor survey as shown previously. This will show the distribution of daylight, the minimum daylight factor and the values at specific work positions.

Using the photoelectric photometer measure the illuminance readings at all work stations and on every work surface, this should be done with the normal workplace lighting switched on, that is, with general and local lamps on. In addtion to this an imaginary grid of one meter squares should be drawn up and illuminance readings taken in the centre of each grid square with both general lighting only on and with general and local on.

To use the photometer:
1 With the cell disconnected adjust the zero as instructed by the maker,
2 connect the cell to the meter,
3 if the cell is separate from the meter and connected by a cable, lay the cell on the surface to be measured and by standing as far from it as possible so as not to overshadow it, read the meter by starting at the highest range setting and reducing the range step by step until the indicator gives an adequate deflection.
4 Correct the readings in accordance with the manufacturers correction code and note the results directly on the plan.

If luminance readings are required of the floor, walls, ceilings and work surfaces then the Universal Photometer should be used as follows.
1 Remove from case and check the meter's mechanical zero and adjust if necessary.
2 Check the battery charge condition as indicated by the makers instructions but this normally involves turning the range finder to any range and depressing a battery check button. If the pointer swings beyond a set mark the battery is in good condition but if the swing is insufficient replace the battery with a good one. Do not attempt to take readings with a low battery as they will be meaningless.

3 Hold the meter up to ones eye and point it at the surface to be measured by viewing through the viewfinder. The surface being measured is that seen within the target circle.
4 Move the control switch to 'Lum. Internal Cell' and turn the meter range switch step by step until an adequate deflection is obtained on the meter. The actual value of luminance is obtained multiplying the meter reading by the range value used.
5 Note the results.

Results and calculations

In addition to recording the results on the plans if further calculations are required then they should be tabulated as shown in tables 6.3, 6.4 and 6.5.

Table 6.3. Recording of illuminance results

Measuring position (as shown in sketch)	Illuminance E (lux)	
	with artificial light only	with artificial and local light
Total		
Mean		

Calculate diversity factor from:

$$DF = \frac{\text{minimum illuminance}}{\text{mean illuminance}}$$

Calculate utilisation factor from:

$$UF = \frac{\text{mean illuminance with artificial light} \times \text{area of room}}{\text{total wattage of lamps in room}}$$

Calculate maintenance factor from:

$$MF = \frac{\text{mean illuminance} \times \text{floor area}}{UF \times \text{total wattage of lamps}}$$

Record luminance in candela m^{-2} in a table similar to table 6.4 and calculate reflectance on the same table.

Table 6.4. Recording of luminance and reflectance results

Surface	Luminance candela/m²	Reflectance = $\dfrac{\text{luminance}}{\text{mean illuminance}}$
Floor		
Walls: 1		
2		
3		
4		
Ceiling		
Horizontal machine surfaces: 1		
2		
3		
etc		
Vertical machine surfaces: 1		
2		
3		
etc		

Record basic room data on a table similar to table 6.5.

Table 6.5. Basic room data sheet

Date and time of survey		Comments on nature of lighting	
Place surveyed (address)		Daylight: side windows	
Room surveyed		roof glazing	
Purpose of survey		Artificial light	
Principle visual tasks		Daylight and artificial	
Planes on which work is done		State of fittings: clean/dirty	
		State of windows: clean/dirty	

Recommended IES Code values:
  illuminance
  daylight factor
		State of floors: clean/dirty	

		State of walls: clean/dirty	

dimensions of room:
  length:
  width
  height
		State of diffusers: clean/dirty	

		State of mountings: clean/dirty	

Windows:
  width
  height
		Lamp/luminaire:
		  types
		  wattage	

		Diffusers/shades:
		  present/absent
		  clean/dirty	

By comparing the results with the standards of workplace illumination recommended in the IES Code, places which require attention can be listed. Improvements can involve either a general cleaning and redecoration of the area or a redesign of the lighting system.

Further reading

The Chartered Institute of Building Services. *The IES Code for Interior Lighting.* CIBS, London, 1977.

Longmore J. Lighting. In Waldron HA & Harrington JM (eds) *Occupational Hygiene.* Blackwell Scientific Publications, 1980.

CHAPTER 7
OTHER HAZARDS

Introduction

There are two classes of hazards that can occur in the workplace which warrant mention but which are sufficiently specialised to remain outside the jurisdiction of most health and safety personnel unless they have undergone special training for their measurement and control. These are: radiation, both ionising and non-ionising, and microbiological hazards. When radiation occurs in a workplace it is normally an integral part of the work process and is present with the full knowledge of all concerned, having the necessary safeguards built in and, in the case of ionising radiation the appointment of a 'competent person' is a legal requirement. However, microbiological hazards usually occur insidiously as an unintentional release of microorganisms from places where they are being handled, such as, research laboratories or where they have formed fugitive colonies breeding in such places as spray humidifiers. The effects of contact with these microorganisms are not noticed until unpleasant symptoms appear in the workforce in sufficient numbers to warrant attention. Unlike dust, gases or heat, in neither case can the above hazards be immediately perceived by the normal human senses therefore no warning signs are noticed or avoiding action taken. Both these topics are covered briefly in this chapter but no instruction sheets are given and, because of their specialist nature, the reader is strongly advised to consult an expert whenever a problem is suspected.

Ionising radiation

Basically this form of radiation occurs in five categories: alpha and beta particles, X-rays, gamma rays and neutrons, each having its own characteristic and penetrating power thus affecting the human body in different ways and requiring different types of material for shielding. Sources of ionising radiation are used in industry for a variety of purposes some of which are as follows: photo-examination of materials, guidance of moving tools, welding, curing, sterilisation, as limit sensors and in chemical analysis. They are also widely used in medicine for: photo-examination, treatment, sterilisation and analysis. Normal sources of ionising radiation are well shielded to prevent unwanted emission thus dangerous worker exposure rarely occurs and then only as a result of careless handling or accidental damage.

In measuring ionising radiation two factors need to be known: radiation intensity as required in radiation surveys and personal exposure or dose as required for workers in daily contact with areas where radiation could be emitted.

Other Hazards

Instruments available

Radiation intensity is measured by two basic instruments: the Geiger-Muller Counter and the Scintillation counter. Both provide a numerical value known as a 'count' but to relate that count to true radiation intensity it is necessary to calibrate the instrument against a known value of intensity.

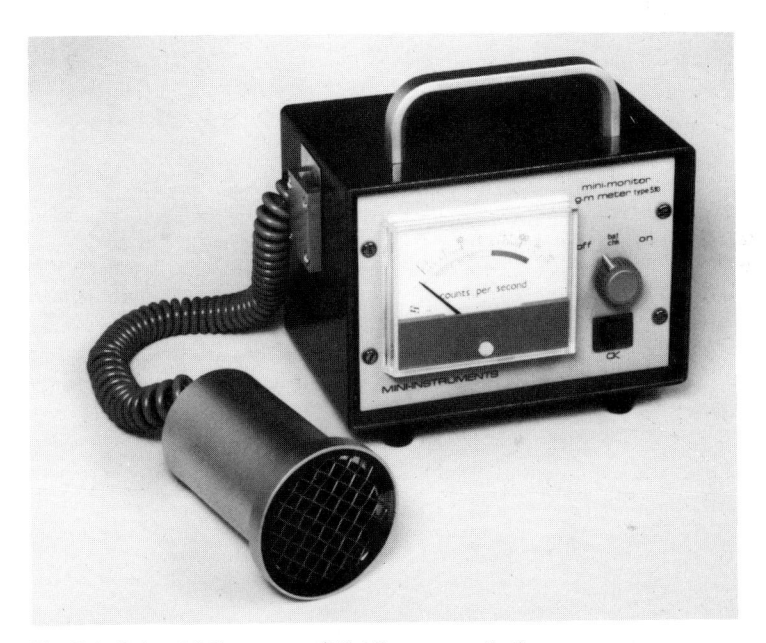

Fig. 7.1. Geiger-Muller counter (Mini-Instruments Ltd).

The Geiger-Muller counter

This instrument measures radiation intensity and consists of a gas filled tube containing a positively charged wire placed concentrically through it, the negative pole being the tube wall. When alpha, beta or gamma radiation enters the tube the gas is ionised and a small electric current flows through it which is amplified and indicated on a meter in counts per second. An audible percussive sound is also provided which increases in pulse as the count rate increases. An example of such an instrument is shown in fig. 7.1.

Chapter 7

The Scintillation counter

This instrument also measures radiation intensity and uses a screen of a material which emits flashes of light when bombarded with alpha, beta, gamma and/or slow neutron radiation. These light flashes are converted to an electric current which increases as the bombardment increases. The current is amplified and indicated in the same way as the Geiger-Muller counter.

Airborne sampler

As with airborne dust, certain radiation can be sampled on a filter paper using a high volume air sampler of the type shown in fig. 1.6 in the chapter on dust. A known volume of air is drawn through a filter which is removed at the end of the sampling period and scanned for its radioactivity using a counter.

Film badge personal dose indicator

A film sensitive to radiation is housed in a specially designed plastic casing containing windows of various materials which shield certain kinds of radiation but which allows others to pass through. The device is worn during periods of exposure, one badge usually lasting a week or longer. The film is sent away to be developed and analysed for the accumulated dose of the various types of radiation. A permanent record of the workers' personal exposure is made.

Thermoluminescent personal dosemeter (TLD)

Some materials such as lithium fluoride can change to what is known as an 'excited' state when bombarded with ionising radiation. This state is reversed only on the application of heat when the crystals return to normal but with a measurable emission of light. Thus a small badge containing these crystals can be used as a dosemeter as the degree of irradiation can be related to the amount of light produced on heating. An advantage with this type of dosemeter is that it is small and its analysis can be quickly and automatically performed.

Non-ionising radiation

The main sources of non-ionising radiation are: ultra-violet and infra-red light, ultrasound, lasers and microwaves.

Emissions of ultra-violet and infra-red light to have a biological effect mainly on the eyes but the wavelength and intensity of radiation are important. Intensity is difficult to measure because the instrument must be sensitive to the wavelength of the emitted radiation. Exposure limits

Other Hazards

for ultra-violet radiation have been specified in the United States as being $100J\ m^{-2}$ for a wavelength of 200nm reducing to $34J\ m^{-2}$ at 280nm and increasing to $10\,000J\ m^{-2}$ at 315nm, thus a wavelength range of 230 to 300nm is the most hazardous.

Ultrasound is used for cleaning and if emitted at a sufficiently high intensity and at the right frequency can cause human damage although no authoritative standards have been set for exposure.

Lasers will cause damage because the energy is concentrated on a very small area and can burn the skin and retina of the eyes if focussed upon them. There is no need to measure intensity of emission as it is usually known or can be calculated from the manufacturers stated power of the device.

Microwaves are used for heating in industry and for cooking in commercial, industrial and domestic premises. It is also emitted from

Fig. 7.2. Microwave leakage monitors (Rohde and Shwarz UK Ltd).

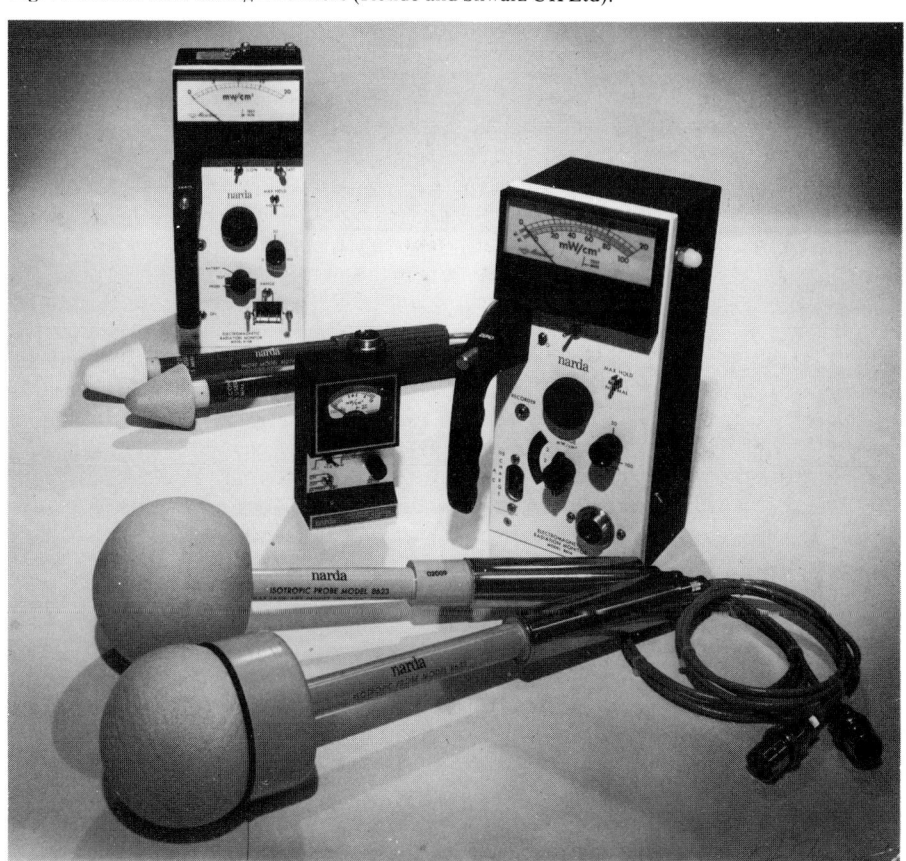

radio and radar transmitters. Its measurement is difficult because microwaves occur over a wide range of wavelengths and the instrument must be sensitive to the wavelength emitted. Microwave ovens for cooking are becoming common and leakage of energy will occur if the doors are improperly sealed or become damaged as a result of wear and tear. Any substance in the path of a fugitive beam will become heated, human flesh being particularly sensitive. Most microwave ovens operate at a frequency of 260MHz and instruments are available to measure leakage around the seals of ovens at this frequency (see fig. 7.2). Standards that apply in the USA but which are often used in Britain is 10mW cm^{-2} but some authorities consider this to be too high. If an emission exceeding 1mW cm^{-2} is measured around the seal of a microwave oven then it must be assumed to require attention.

Microbiological hazards

There is a multitude of microorganisms which can be airborne in work situations in the form of bacteria, viruses, spores, rickettsiae and protozoa. The many sources include people, animals, laboratories using pathogenic organisms, kitchens and places where water stagnates at the correct temperature for incubation and breeding. Therefore virtually every person at work, at home and elsewhere can be exposed. In most cases human defence mechanisms will cope with the exposure but a continued contact with a microorganism or the exposure to a virulent strain can produce an adverse reaction.

Measurement of exposure involved firstly the sizing of the organism, secondly its identification and thirdly its airborne concentration. Because of the very wide range of organisms that occur, for example, over 30 different types have been identified in the water from one humidifier, it requires the work of an expert microbiologist to sample, breed and examine under the microscope the airborne organisms to identify them and to establish their airborne concentration. It is beyond the scope of this book to cover such a wide subject, thus if it is suspected that workers' symptoms are caused by airborne microorganisms then a microbiologist should be consulted.

Further reading

International Commission on Radiological Protection. *ICRP 26.* Published by ICRP through Pergamon Press, 1977.

Radiation Protection Instrumentation and its Application. *Report 20.* International Commission on Radiation Units and Measurements, Washington DC, 1971.

Doran D. Ionising radiation. In Waldron HA & Harrington JM (eds) *Occupational Hygiene.* Blackwell Scientific Publications, 1980.

Kanagasaby S. Non-ionising radiation. In Waldron HA & Harrington JM (eds) *Occupational Hygiene.* Blackwell Scientific Publications, 1980.

The Ionising Radiations (Sealed Sources) Regulations, 1969. HMSO, 1969.
The Ionising Radiations (Unsealed Radioactive Substances) Regulations, 1968. HMSO, 1968.
Microbiological hazards. In Allen RW, Ells MD & Hart AW, *Industrial Hygiene.* Prentice Hall, 1976.

CHAPTER 8
SURVEYS

Introduction

The foregoing chapters have indicated the methods of taking workplace environmental measurements and the instruments available to do so but it is important to realise that to obtain a true picture of the health hazards of a working environment no single measurement will suffice. This is because workplace pollution rarely occurs evenly spread in concentration or intensity over the whole workplace or over the whole working period. In the case of the emission of a gas or particles of dust the concentration is greatest at the point of emission but it may fluctuate as the process progresses. As the pollutant moves away its behaviour and dispersion will depend upon the air currents occurring in the room and therefore will vary with the movement of people, machinery and both naturally and mechanically induced air currents. Even one shift may not be typical of others in the same place due to the variability and cyclic nature of most processes.

Therefore surveys must be planned within the resources available to obtain the best possible information on the hazards as they affect individual workers and on the workforce as a whole. Generally speaking the more measurements that are taken over the longest period of time the more valuable the results will be. If important decisions are to be made or expensive equipment purchased based upon the results of measurements then they must be taken as accurately and as scientifically as possible and preferably by a professional occupational hygienist.

Planning

It is important to plan in advance any survey, hence a visit to the measuring sites should be made wherever possible. If the site is too far distant to make a visit then a plan of the areas to be surveyed should be obtained and these must show the position of the main emitters of pollutants, heat, noise or radiation together with the indication of where the operators are stationed.

A full compliment of equipment is required at the site before starting. To have to return to base for some forgotton item is, at best, time wasting and, at worst, is impossible for reasons of distance. One or two days of measuring time can be lost when far from home base and some vital item is missing. Check lists are given for survey items and are offered as a guide. They should be added to if special items are required and in the light of experience. It may not be necessary to include every item on each list but it is better to consider and reject rather than not to consider at all.

A photograph of the workplace is an invaluable aid to memory, particularly when it is inconvenient to return to the site. Therefore a camera has been included in all check lists but permission must always be sought to use it.

Labour and assistance

It is always useful to plan a survey involving two people one of whom need not be involved in the technical side but should be available to record results and note the workplace operations that are taking place. A single surveyor can easily become harassed if too much work is attempted and under such pressure vital information is not recorded and unrepeatable data are lost. A good surveyor will come with record sheets already prepared with columns for every item to be noted and as each reading is taken will note its value in the appropriate place. The amount of work that can be done in a shift will depend upon the layout of the work area to be surveyed but it is wise not to undertake too much, for example, in the case of an airborne dust survey, no person should be expected to have more than ten sampling pumps running simultaneously without the help of an assistant and if the area to be covered involves several workrooms then even less than this should be planned.

The places where timed sampling is being carried out should be supervised not only to make a note of the type of work tasks that are being carried out but to watch the equipment to minimise the risk of pilfering and to ensure that the results are not being adversely affected by deliberate over or under exposure. It has been the authors' unfortunate experience occasionally to have a handful of dust thrown across a sampling filter or a charcoal tube placed above the surface of a solvent emitter to deliberately increase the concentration measured. The opposite sometimes occurs where the sampling device is placed in a fresh air situation to ensure that a low concentration is obtained. The continued presence of the surveyor will minimise the occurrence of such deliberate action.

Results

It is not within the remit of this book to give advice on the interpretation of results but standards have been referred to under the various operations where applicable. Standards for workplace environments are published by various sources some of which are listed in further reading but it must be remembered that standards are offered as a guide and not as the strict dividing line between what is hazardous and what is not. Wherever possible action should be taken to improve the situation if levels of more than half the recommended standard is measured as this indicates that conditions are approaching that standard. Also there may be persons present in the workplace who are more susceptible to the pollutant than the majority for whom the standard was chosen.

When all the results of a particular survey are available it is wise to examine them carefully to establish how representative of the workplace they really are. Every possible confounding factor should be considered and places and times when errors could have occurred in the measuring technique should be noted. It is only when one considers how unrepresentative of a workplace even the most carefully planned and executed surveys can be, that it is realised how unwise it is to stick rigidly to the TLV's or other criteria and standards.

Further reading

Planning surveys and use of results. In Jones AL, Hutcheson DMW & Dymott SM, *Occupational Hygiene, an Introductory Guide.* Croom Helm, 1981.

Lee GL, Sampling: principles, methods, apparatus, surveys. In Waldron HA & Harrington JM (eds) *Occupational Hygiene.* Blackwell Scientific Publications, 1980.

Oakes D, Statistics. In Waldron HA & Harrington JM (eds) *Occupational Hygiene,* Blackwell Scientific Publications, 1980.

American Conference of Government Industrial Hygienists. *Threshold Limit Values for Chemical Substances and Physical Agents in the Workroom Environment.* ACGIH, Cincinnati, Ohio, published annually.

International Labour Office. *Occupational Exposure Limits for Airborne Toxic Substances.* ILO, Geneva, 1977.

Dust survey equipment check list *Surveys*

Item as required	Amount	Packed	Returned
Adhesive tape			
Camera and film			
Carrying case			
Clip board			
Cyclone filter holder			
Dust lamp			
Electronic calculator			
Filters (weighed)			
Filter holders and covers			
Forceps or tweezers			
Harnesses			
Knife			
Labels			
Membrane filters			
Paper			
Pens and pencils			
Petri-slides			
Petri-dishes			
Plastic bags			
Pumps—high flow			

continued overleaf

Chapter 8 **Dust survey equipment check list** *continued*

Item as required	Amount	Packed	Returned
Pumps—medium flow			
Results sheets			
Rotameter and calibration chart			
Safety pins			
Scissors			
Screwdrivers (small)			
Smoke tube kit			
String			
Tape measure			
Tripods			
Tubing			
Timer or stop watch			
Safety clothing: shoes			
goggles			
helmet			
respirator			
gloves			
overalls			
ear defenders			

Gases and vapours survey equipment check list　　　　　*Surveys*

Item as required	Amount	Packed	Returned
Adhesive tape			
Adsorbent tubes			
Bubblers (empty)			
Bubblers (full)			
Bubbler reagent			
Camera and film			
Carrying case			
Clip board			
Colourimetric detector tubes and pump			
Direct reading instrument			
Electronic calculator			
Filters			
Filter holders			
Forceps or tweezers			
Harnesses			
Knife			
Labels			
Passive samplers			
Paper			
Pens and pencils			
Plastic bags			
Pumps—low flow			

continued overleaf

Chapter 8   **Gases and vapours survey equipment check list** *continued*

Item as required	Amount	Packed	Returned
Pumps—medium flow			
Result sheets			
Rotameter and calibration chart			
Safety pins			
Sampling bags: mylar or tedlar			
Scissors			
Screwdriver (small)			
Smoke tube kit			
String			
Tape measure			
Tripods			
Tubing (check bore)			
Vacuum tubes			
Timer or stop watch			
Safety clothing: shoes			
goggles			
helmet			
respirator			
gloves			
overalls			
ear defenders			

Thermal survey equipment list

Surveys

Item as required	Amount	Packed	Returned
Adhesive tape			
Aspirated psychrometer			
Botsball thermometer			
Camera and film			
Carrying case			
Clip board			
Distilled water			
Dry bulb thermometers			
Electronic calculator			
Globe thermometer (large) and charts			
Globe thermometer (small) and charts			
Heat index charts			
Kata thermometer and charts			
Natural wet bulb thermometer: wicks and beakers			
Paper			
Pens and pencils			
Psychrometric charts			
Reflective foil and corks			
Results sheets			

continued overleaf

Chapter 8 **Thermal survey equipment list** *continued*

Item as required	Amount	Packed	Returned
Sling psychrometer (whirling hygrometer)			
Smoke tube kit			
Spare thermometers			
Spare wicks			
Tape measure			
Thermos flask and hot water			
Timer or stop watch			
Tripod: stand			
boss heads			
clamps			
WBGT meter			
Safety clothing: shoes			
goggles			
helmet			
respirator			
gloves			
overalls			
ear defenders			

Ventilation survey equipment check list

Item as required	Amount	Packed	Returned
Anemometer, heated head type			
Anemometer, vane type			
Aneroid barometer			
Camera and film			
Calibration charts			
Carrying case			
Clip board			
Desk fan			
Diaphragm pressure gauge			
Drill and bit for hole boring in ducts			
Electronic calculator			
Log/linear graph paper			
Manometers			
Manometer liquid			
Marker pen			
Paper			
Pen and pencils			
Pitot-static tubes			

continued overleaf

Chapter 8 **Ventilation survey equipment check list** *continued*

Item as required	Amount	Packed	Returned
Plasticine or 'blue tack'			
Plugs for holes in ducts			
Results sheets			
Smoke tube kit			
Tape measure			
Thermometer			
Timer or stop watch			
Tracer gas			
Tracer gas detector			
Tubing for gauges and manometers, blue and red			
Safety clothing: shoes			
goggles			
helmet			
respirator			
gloves			
overalls			
ear defenders			

Noise survey equipment check lists *Surveys*

Item as required	Amount	Packed	Returned
Batteries, spare for meters			
Calibrator			
Camera and film			
Carrying case			
Clip board			
Dosemeters			
Electronic calculator			
Microphones for meters			
Microphone extensions for meters			
Noise rating curves			
Octave band analyser			
Octave band charts			
Paper			
Pens and pencils			
Pistonphone			

continued overleaf

Chapter 8 **Noise survey equipment check lists** *continued*

Item as required	Amount	Packed	Returned
Results sheets			
Screwdriver (small)			
Sound level meters			
Tape measure			
Tape recorder, connections and tapes			
Tripods			
Timer or watch			
Safety clothing: shoes			
goggles			
helmet			
respirator			
gloves			
overalls			
ear defenders			

Lighting survey equipment check list *Surveys*

Item as required	Amount	Packed	Returned
Calibration charts			
Camera and film			
Carrying case			
Clip board			
Daylight factor meter			
Electronic calculator			
Graph paper			
Hagner Universal Photometer			
IES Code			
Measuring tape (10m)			
Munsel charts			
Paper			
Pens and pencils			
Photometer			
Results tables and plans			
Screwdriver (small)			
Tape measure			
Safety clothing: shoes			
goggles			
helmet			
respirator			
gloves			
overalls			
ear defenders			

Chapter 8   Suppliers of equipment

Acoustic calibrators: *see* Sound level meters
Adsorbent tubes: Casella London Ltd, D.A. Pitman Ltd
Anemometers and airflow meters: Abbot Birks Ltd, Airflow Developments Ltd, BIRAL(TSI), Prosser Scientific Ltd
Adhesive tape (Sleek strapping): Smith and Nephew Ltd
Aspirated psychrometer: Casella London Ltd
Balances: Oertling Ltd, A. Gallenkamp & Co Ltd
Botsball thermometer: *see* Thermometers
Bubblers: Casella London Ltd, A. Gallenkamp & Co Ltd
Carrying cases: Topper cases
Colourimetric gas detectors: Draeger Safety Ltd, D.A. Pitman Ltd, Sabre Gas Detection, Vinton Instruments Ltd.
Cyclone filter holders: *see* Filter holders
Daylight factor meter: Diffusion Systems Ltd
Diaphragm pressure gauges: Granville Controls Ltd
Direct reading dust instruments: Analysis Automation Ltd, BIRAL(TSI), Gelman Sciences (Royco), MDA Scientific (UK) Ltd, Rotheroe and Mitchell Ltd
Direct reading gas instruments: Analysis Automation Ltd, Foxboro Analytical, Kemtronics (UK) Ltd, MDA Scientific (UK) Ltd, D.A. Pitman Ltd, Rotheroe and Mitchell Ltd, Sabre Gas Detection, Shaw City Ltd
Dust Lamp: A & G Marketing Ltd
Filters: A. Gallenkamp & Co Ltd, Gelman Sciences (UK) Ltd, Millipore (UK) Ltd, Sartorius-Instruments Ltd, Whatman Laboratory Products Laboratory
Filter Holders: Casella London Ltd, Gelman Sciences (UK) Ltd, Rotheroe and Mitchell Ltd
Forceps: A. Gallenkamp & Co Ltd
General laboratory equipment: A. Gallenkamp & Co Ltd
Globe thermometers: *see* Thermometers
Hagner Universal Photometer: Hagner International (UK) Ltd
Harnesses: Casella London Ltd, Rotheroe and Mitchell Ltd
Kata thermometer: *see* Thermometers
Manometers: Airflow Developments Ltd
Membrane filters: *see* Filters
Microphones: *see* Sound level meters
Microscopes: A. Gallenkamp & Co Ltd (Olympus), Vickers Instruments
Microscope dispersion staining objectives: MaCrone Research Associates Ltd
Microscope eyepiece graticules: Graticules Ltd
Microscope slide mounting fluids: McCrone Research Associates Ltd
Microscope slides and cover glasses: A. Gallenkamp & Co Ltd
Nose dosemeters: Bruel and Kjaer (UK) Ltd, Computer Engineering Ltd, Cirrus Research Ltd, Dawe Instruments Ltd, DuPont (UK) Ltd, General Acoustics Ltd, Pulsar Instruments, Reten Acoustics Ltd
Passive sampler: D.A. Pitman Ltd, Vinton Instruments Ltd, Draeger Safety Ltd, Du Pont, Perkin-Elmer Ltd
Paper tape gas detectors: MDA Scientific (UK) Ltd
Petri dishes and Petri slides: A. Gallenkamp & Co Ltd, Millipore (UK) Ltd
Photometers: Hagner International (UK) Ltd, Salford Electrical Instruments Ltd

Pistonphones: Bruel and Kjaer (UK) Ltd, General Acoustics Ltd
Pitot-static tubes: Airflow Developments Ltd
Pressure gauges: Airflow Developments Ltd, Granville Controls Ltd
Pumps for air sampling: Casella London Ltd, DuPont (UK) Ltd, MDA Scientific (UK) Ltd, Rotheroe and Mitchell Ltd, Vertec Scientific
Radiation monitors: Mini-Instruments Ltd, Rohde and Schwarz (UK) Ltd (Narda), Vinton Instruments Ltd, Vertec Scientific
Rotameters: Casella London Ltd, A. Gallenkamp & Co Ltd, G.A. Platon Ltd
Sampling bags: Casella London Ltd, Thompson Laboratories
Sling psychrometers: Airflow Developments Ltd, Casella London Ltd
Smoke tube kits: Draeger Safety, MSA (Britain) Ltd
Sound level meters: Bruel and Kjaer (UK) Ltd, Computer Engineering Ltd, Cirrus Research Ltd, Dawe Instruments Ltd, DuPont (UK) Ltd, General Acoustics Ltd, Pulsar Instruments, Reten Acoustics Ltd
Tape recorders: Bruel and Kjaer (UK) Ltd, Hayden Laboratories Ltd (Nagra/Kudelski), Photo Acoustics Ltd (Uher), Racal Recorders Ltd
Thermometers: Casella London Ltd, A. Gallenkamp & Co Ltd, Kane-May Ltd
Tracer gas krypton K^{85}: Amersham International Ltd
Tripod stands: Bowens of London
Tubing (plastic flexible): coloured: Airflow Developments Ltd; clear: A. Gallenkamp & Co Ltd
Vacuum sampling tubes: Casella London Ltd
WBGT meters: Casella London Ltd, Light Laboratories, Vertec Scientific
Wind tunnel (open jet): Airflow Developments Ltd

Addresses

Abbott Birks Ltd
Eastway Portway, Andover
Hants SP10 3NJ
tel: 0264 63261

A & G Marketing
Bridle House, Brent Pelham
Buntingford
Herts
tel: 027978 444

Airflow Developments Ltd
Lancaster Road
High Wycombe
Bucks HP12 3QP
tel: 0494 25252

Amersham International Ltd
White Lion Road
Amersham
Bucks HP7 9LL
tel: 02404 4444

Analysis Automation Ltd
Southfield House
Eynsham
Oxford OX8 1JD
tel: 0865 881888

Bowens of London Ltd
Royalty House
72 Dean Street
London W1
tel: 01 439 1781

BIRAL (Bristol Industrial and
  Research Associates Ltd
PO Box 2, 6 Combe Road
Portishead
Bristol BS20 9JB
tel: 0272 847767

British Standards Institution
2 Park Street
London W1A 2BS
tel: 01 629 9000

Bruel and Kjaer (UK) Ltd
Cross Lances Road
Hounslow TW3 2AE
tel: 01 570 7774

Casella London Ltd
Regent House
Britannia Walk
London N1 7ND
tel: 01 253 8581

Chartered Institute of Building
  Services
Delta House
222 Balham High Road
London SW12 9BS
tel: 01 675 5211

Cirrus Research Ltd
1–2 York Place
Scarborough
Yorks YO11 2NP
tel: 0723 71441

Computer Engineering Ltd
Wallace Way
Hitchen
Herts SG4 0SE
tel: 0462 52731

Dawe Instruments Ltd
Concord Road
Western Avenue
London W3 0SD
tel: 01 992 6751

Du Pont (UK) Ltd
Du Pont House
18 Bream's Buildings
London EC4A 1HT
tel: 01 242 9044

Diffusion Systems Ltd
43 Rosebank Road
London W7 2EW
tel: 01 579 5231

Draeger Safety
Draeger House
Sunnyside House
Chesham
Bucks HP5 2AR
tel: 0494 74481

Foxboro Analytical
28 Heathfield
Stacey Bushes
Milton Keynes
Bucks MK12 6HR
tel: 0908 318222

Addresses

A. Gallenkamp and Co Ltd
PO Box 290
Technico House
Christopher Street
London EC2P 2EP
tel: 01 247 3211

Gelman Sciences Ltd
10 Harrowden Road
Brackmills
Northampton NN4 0EB
tel: 0604 65141

Granville Controls
63 Pritchett Street
Aston
Birmingham B6 4EX
tel: 021 359 8207

Graticules Ltd
Morley Road
Botany Trading Estate
Tonbridge
Kent TN9 1RN
tel: 0732 35906

Hagner International (UK) Ltd
42 Little London
Chichester
Sussex PO19 1PL
tel: 0243 781290

Hayden Laboratories Ltd
Hayden House
Chiltern Hill
Chalfont St Giles
Gerrards Cross
Bucks SL9 9UG
tel: 02813 88447

Kane-May Ltd
Burrowfield
Welwyn Garden City
Herts AL7 4TU
tel: 07073 31051

Kemtronix (UK) Ltd
Aldworth Road
Compton
Berks RG16 0RD
tel: 063 522 470

Light Laboratories
10 Princes Street
Brighton BN2 1RD
tel: 0273 27666

McCrone Research Associates Ltd
2 McCrone Mews
Belsize Lane
London NW3 5BG
tel: 01 435 2282

MDA Scientific (UK) Ltd
Ferndown Industrial Estate
Unit 6
Haviland Road
Wimborne
Dorset BH21 7RF
tel: 0202 874318

Millipore (UK) Ltd
11/15 Peterborough Road
Harrow HA1 2YH
tel: 01 864 5499

MSA (Britain) Ltd
East Shawhead
Coatbridge
Scotland ML5 4TD
tel: 0236 24966

Mini-Instruments Ltd
8 Station Industrial Estate
Burnham-on-Crouch
Essex CM0 8RN
tel: 0621 783282

Narda (Rohde and Schwarz UK Ltd)
Rohde and Schwarz House
Roebuck Road
Chessington
Surrey KT9 1LP
tel: 01 397 8771

Oertling Ltd
Cray Valley Works
St Mary Cray
Orpington
Kent BR5 2HA
tel: 0689 25771

Perkin-Elmer Ltd
Post Office Lane
Beconsfield
Bucks H99 1QA
tel: 049 46 6161

Photo Acoustics Ltd
58 High Street
Newport Pagnell
MK16 8AQ
tel: 0908 610625

Addresses

D.A. Pitman
Jessamy Road
Weybridge
Surrey KT13 8LE
tel: 0932 46327

G.A. Platon
Wella Road
Basingstoke
Hants RG22 4AQ
tel: 0256 26661

Prosser Scientific Instruments
Lady Lane Industrial Estate
Hadleigh
Suffolk IP7 6BQ
tel: 0473 3005

Pulsar Instruments
40/42 Westborough
Scarborough
North Yorks YO11 1UN
tel: 0732 71351

Rotheroe and Mitchell Ltd
Victoria Road
Ruislip
HA4 0YL
tel: 01 422 9711

Reten Acoustics Ltd
78 Bridge Street
Newport
Gwent NPT 4AQ
tel: 0633 59910

Sabre Gas Detection Ltd
Ash Road
Aldershot
Hants GU12 4DD
tel: 0252 316611

Salford Electrical Instruments Ltd
Peel Works
Barton Lane
Eccles
Manchester M30 0HL
tel: 061 789 5081

Sartorius-Instruments Ltd
18 Avenue Road
Belmont
Surrey SM2 6JD
tel: 01 642 8691

ShawCity Ltd
Unit 2
Pioneer Road
Faringdon
Oxfordshire SN7 7BU
tel: 0367 21675

Racal Recorders Ltd
Hardley Industrial Estate
Hythe
Southampton
Hants SO4 6ZH
tel: 0703 843265

Smith and Nephew
Bessemer Road
Welwyn Garden City
Herts AL7 1HF
tel: 07073 25151

Thompson Laboratories
The Stocks, Cosgrove
Milton Keynes
Bucks ML19 7JD
tel: 0908 562925

Topper Cases
St Peters Hill
Huntingdon
Cambs PE18 7ET
tel: 0480 57251

Vertec Scientific
436 Bath Road
Chippenham
Slough
SL1 6BB
tel: 06286 4808

Vickers Instruments
Haxby Road
York YO3 7SD
tel: 0904 31351

Vinton Instruments Ltd
Jessamy Road
Weybridge
Surrey KT13 8LE
tel: 0932 46327

Whatman Laboratory Products
Springfield Mill
Maidstone
Kent ME14 2LE
tel: 0622 61681

List of occupational hygiene consultants *Addresses*

Birmingham

Department of Occupational Health and Safety
The University of Aston in Birmingham
Gosta Green
Birmingham B4 7ET
Contact: Mr M. Piney, tel: 021 359 3611 Ext. 577

Institute of Occupational Health
The Medical School
University of Birmingham
Birmingham B15 2TJ
Contact: Mr F.S. Gill, tel: 021 472 1301 Ext. 3500

Cardiff

Occupational Health and Safety Group
Department of Epidemiology and Community Medicine
Welsh National School of Medicine
Heath Park
Cardiff CF4 4XN
Contact: Mr A. Samuel, tel: 0222 755944

Dundee

Environmental Health Service
Wolfson Institute of Occupational Health
University of Dundee
Level 5
Medical School
Ninewells
Dundee
Contact: Mr T. Gillanders, tel: 0382 644625

Essex

National Occupational Hygiene Service Ltd
49 Viking Way
Pilgrim's Hatch
Brentwood
Essex CM15 9HY
Contact: Mr K.L. Knight, tel: 0277 224183

London

Information and Advisory Service
TUC Centenary Institute of Occupational Health
London School of Hygiene and Tropical Medicine
Keppel Street
Gower Street
London WC1E 7HT
Contact: IAS Co-ordinator, tel: 01 636 8636 Ext. 390

Institute of Environmental Science and Technology
Polytechnic of the South Bank
Borough Road
London SE1 0AA
Contact: Mr C. Money, tel: 01 928 8989 Ext. 2113

Addresses

Manchester

National Occupational Hygiene Service Ltd
12 Brook Road
Fallowfield
Manchester M14 6UH
Contact: Mr E. King, tel: 061 224 2332

Milton Keynes

Thompson Laboratories
The Stocks
Cosgrove
Milton Keynes MK19 7JD
Contact: Miss N. Thompson, tel: 0908 562925

Newcastle upon Tyne

North of England Industrial Health Service
Department of Occupational Health and Hygiene
The Medical School
The University of Newcastle upon Tyne
Newcastle upon Tyne NE14 7RU
Contact: Dr J. Steel, tel: 0632 28511

Analytical services

In addition to the consultants listed above who have their own analytical services the following also offer this facility.

Birmingham

Dutom Meditech Ltd
Warwick Street
Birmingham B12 0NH
Contact: Dr J.M. Thompson, tel: 021 771 3000

Guildford

Wolfson Bioanalytical Unit
Institute of Industrial and Environmental Safety
University of Surrey
Guildford
Surrey GU2 5XH
Contact: Dr E. Reid, tel: 0483 71281

Engineering control consultant and contractor

Portsmouth

O'HEAL
Tadgell Mount
Catherington
Portsmouth PO8 0TB
Contact: Mr D. Asker-Brown, tel: 0705 594573

INDEX

Acoustics 94
Acoustical Engineer 102
Aerodynamic diameter 1
Air analysis, direct and indirect, 29
  changes per hour 70, 90–1
  clean 70
  cooling power 69, 75
  dirty 70
  dust laden 5
  flow patterns 78, 87
  flow rates 2–5, 14, 17–27, 33–4, 39–45, 70, 83, 89
  sampling 13–26, 38–45
  ventilation 83–90
  flow splitters 88
  humidity 16, 73
  pressure 70
  speeds 70, 90
  stream lines 87–9
  temperature 82–3
  velocity 52–7, 68–9, 74–88
    instruments 78, 83
Airborne
  asbestos 21–2
  chemicals 30
  concentration 32–41
  contaminant 3
  dust 3, 32
  fibre counts 12
  gas 41
  particles 7
  pollutants 32, 34
Air flow 63
  mechanical 70, 89
  meters 57
  natural 70
  patterns 78, 87
Aluminium foil 63
Alveoli 2
American Conference of Government Industrial Hygienists 62
Ampoule 47
Analysis 3, 7, 30
  chemical 29
  methods of 26
Analytical instruments 30
Analytical services 147–8
Anemometers 89–92
  heated head 74, 82
  vane 74–80, 87
Asbestos 21–2
  control limits for 25

  microscopic examination of 24
  regulations 24–5
  types 25
Aspirated psychrometer 54–5, 66
Asphyxia 29
Asthma 1
Atmosphere, flammable 75–6
Atomic adsorption spectroscopy 26

Bags, sampling 42–3
Balance
  micro 8
  sensitivity of 17
Barometer 91
  aneroid 79
Battery 7, 12
  chargers 7
  test 100
Battery powered fan 54
Beaker 63
Bel 95
Bellows pump 45–7
  checks of 47
Beta-ray absorption 8, 12
Botsball 59
Breathing zone 15, 38–9, 43
  personal exposure 29–30
British Standards 80–2, 98
BS848 80–2
Bronchi, bronchioles, bronchitis 1
Bubbler 32, 34, 45
  fritted 32
  sampling 43
  spill proof 43
Burette 12, 15

Calibration 11–12
  anemometer 91
  chart 14, 93
  devices 78
  filter 12
  rotameter 2, 11, 15
  soap bubble method 11–15
  sound level meter 99, 100, 103
  tone 100
Camera 116, 129
Carcinoma 1
Carbon monoxide 30, 37
Cell
  light sensitive 37, 111–13
Charcoal 30, 34, 37

149

Index

Clearing
  fluids 22–4
  membranes 22
Chronic lung disease 1
Ciliary movement 1
Central nervous system 29
Coal 2
  dust 26
Cold, extreme 53
Collection
  bags 30, 41
  devices 30
Colourimetric detectors 29, 35, 45–51
  kits 36, 45
  long term 30, 45, 50
  shelf life 36
  staining 48–51
Colour
  correction factors 111
  rendering of lamps 117–18
Concentration
  airborne fibres 22, 25
  airborne gas 44–5
  localised peaks of 8
  peaks and troughs of 29, 41
  time weighted average 8, 29, 37–43
  total dust 15, 19
Conduction 52
Control limits 25
Convection 52
Cooling
  power 64
  time 65
Cotton wick 53–4
Critical orifice 34
Counting
  automatic 12
  fibre 3, 21
  inter-laboratory 26
  inter-person 26
  size selective 12
Cyclone 3, 5, 19–21

Daylight 114, 117
  factor 110, 117–18
    contours 116
    meter 112, 115
    measurement of 113
Decibel 95, 98, 101
  A-weighting 103–6
  weighting 96, 100
  addition of 95
Density
  air 59, 71, 77, 93
  liquid 71
Desorption 30
  thermal 35
Detector tubes 45–51
Diffusion 32
Digital counters 1, 22
  display 8, 12, 79, 107
  indicator 74

Direct reading instruments
  for dust 2, 7–12
  for gases and vapours 37, 41, 43, 87
Discharge nozzle 41–2
Distilled water 53–4, 63, 66
Diversity factor 110, 119
Draughts, cross 87, 89
Duct
  circular 80–2
  pressures 71
  rectangular 80–3
  wall 80, 82
Ducting 70–93
  dampers 84
  fitting etc, 70
Dust 1–28
  collector
  concentration 2, 15, 19
  mass of 2, 8, 12, 20
  nuisance 1, 26
  respirable 19, 20
  size 1, 12
  specific 3, 26
  survey check list 131–2
  total 3, 15, 19, 20

Ear 94, 96
Elutriator 5
  parallel plate 5
  vertical 12
Emissivity 52
Enclosed vessels 45
Engineering Control Consultant and
    Contractor 148
Enthalpy 59, 62
Equivalent noise energy 101
Evacuated vessels 30
Evaporation 52
Extraction
  enclosures, hoods and fume
    cupboards 70, 86
  slots 70, 86, 88
Extract ventilation 26
Eye
  fatigue 109
  radiation effects on 124
  retina 125
  strain 109

FAM 12
Fans 70, 84–5, 89, 92
Fan
  blades 80
  desk 90
Fibre
  count 12, 21–5
  definition of respirable 24
Fibres
  asbestos 21–2, 26
  concentration of 25
  glass 3
  health hazard and inhalation of 22
  man made 26

Index

Fibrous 3
Film badge 124
Filters
  choice of 26–7
  clearing 22–4
  control 16–19
  damaged 16, 19
  dust 3, 22
  gridded 3, 22
  membrane 11, 22
  open face 22
  overloaded 19
  paper 3, 124
  pore sizes 3
  types 3, 26
  ventilation 70, 84
Filter holders 3–5, 15–20
  choice of 26
Filtration
  methods 8
  system 2
Flicker 117–8
Flow meter
  air 77
  calibrated 39
  soap bubble 34
Flow rate
  air 2–5, 14, 17–27, 33–4, 39–45, 70, 83, 89
  pulsating 6, 15, 84
  sampling 12–27, 39–45
  smooth 5, 19
  steady 12
  too high 19
  ventilation 83–90
Flow smoother 6, 15, 20
Fluorescent spectroscopy 26
Forces, centrifugal, electrostatic and inertial 3
Frit 43
Fumes 2, 3
  caused by welding 26

Gas chromatograph 37
Gases and vapours 11, 29–51
  measurement of 45
  survey check list 133
Gauge
  diaphragm 73–4, 82, 84
  liquid filled 73, 86
  Magnehelic 73
  pressure 70, 76, 82
Geiger–Muller
  counter 90, 123, 124
  tube 89
Glare 109
  discomfort and disability 117
Globe thermometer 52–59, 66
  chart 69
Grab sample 29–32, 45
Graticules 22–3
Gravimetric analysis 26
Grid 3, 22, 87–8

Hard rocks 2
Harness 2, 3, 8, 15, 16, 37, 43
Hagner Universal Photometer 113
Health and Safety Commission 25
Health and Safety Executive 22
  standards for asbestos 25
Heat 52–69
  exchange, flow, transfer 52
  extremes of 53
  indices 60, 62, 65
  loss 89
  radiant 54
Heaters 70
Hearing
  damage 100–2
  defenders 102
  frequency 94
  threshold of 95
Heavy metals 26
Humidity 15, 53–5, 63
Humidifier 70
  spray 122
Hydrocarbons 30

IES Code 117
Illumination 113–14
  artificial 109, 113–14, 117–19
  levels of 116–8
  natural 109, 113–14, 117–19
Illuminance 110, 117–19
Incubate 126
Indices
  heat 60, 62, 65
  heat stress 62
Infra-red
  analysis 26
  beam 37
  instrument 37–8
  radiation 124
Integrating instruments 57
  heat stress meter 60
International standards 98
Intrinsically safe 75
Ionising radiation 122–4
Irradiation 124

Kata thermometer 57–9, 63–4, 75
  bulb 66
  chart 67
  factor 57, 63–69
  range 64
Kinetic energy 71

Labour and assistance 129
Lamp, dust 27–8
Lamps
  discharge 117
  fluorescent 117
Lasers 124–5
L_{eq} 96, 105–6
  measurement of 105
L_{10}, L_{90} 96
Lead, 26

151

Index

Lighting 109–121
  adequacy of 117
  levels 109
  measurement of 118
  sources of 111
  suitable 117
  survey 116
    check lists 141
  units of 109–112
  workplace 109
Light scattering 10, 12
Liquid crystal 54, 107
Log linear and log Tchebycheff rules 80–2
Luminance 110–12, 118–19
  results 120
Luminous intensity and flux 109
Lung 2

Maintenance factor 111, 119
Manometer 71, 73, 84, 91–2
  liquid filled 82, 85–6
  portable inclined 72–3, 82
Mass 20
Mean radiant temperature 52, 66, 69
  calculation of 65
  definition of 52
  nomogram 68
  use of chart 69
Measurement of
  airborne asbestos 22–3
  airborne concentration of gas 45
  airflow in ducts 79
  daylight factors 113
  lighting 118
  natural infiltration 89
  noise spectrum 102
  octave bands 102
  suction inlet performance 86
  thermal environment 62
  total airborne dust 15
Mercury vapour 37
Metal fume fever 2
Methane 30
Micro-ammeter 112
Microbiological hazards 122, 126
Microorganisms 122, 126
Microphone 98, 100–3, 106, 108
Microscope
  examination for asbestos using 24
  fields of view 22
  magnification of 24
  optical 3, 22
  scanning electron 3
Microscopy 22, 25–6
Micro-wave
  cooking 125
  leakage monitor 125
  ovens 126
Mines 5
Miran infra-red analyser 38
Mists 2, 3
  oil 26

Moisture content 52, 59, 62, 66
  atmospheric 3
Molten metal 2
MRE 113 5, 9, 26
Mutagenesis 29
Mylar film 10

Narcotic 29
Nitrogen, oxides of 30
Noise 94–108
  attenuation 102
  background 101
  calibrators 100–1
  dose 94, 96
  dose meters 98–9, 106–8
  exposure 94, 101
  frequency 94, 96
  health hazard 94
  impact 96
  impulse 98
  instrumentation 94
  levels 94, 104
  measurements 98, 100
    personal 98
  rating curves 103–5
  recommended ratings 105
  stress 100–2
  survey check list 139
Non-ionising radiation 124
Nuisance dust 1

Occupational Hygiene, list of
  consultants 147–8
Occupational Hygienists 128
Octaveband
  analyser 97, 100
  analysis 102
  filter set 103
  mid-frequencies 96, 104
  third 102
Orifice 39, 49
  plate 9, 92–3
  suction pump 47, 51
Oxygen transfer 1

Paint solvents 40
Paper tape monitor 29, 37–8
Particle
  count 3, 10, 12
  mass 21
  size 21
Passive samplers 32–3
Pathogens 80, 126
Peizo-electric microbalance 8, 12
Percentage saturation 59, 62
Permeation 32
Personal exposure
  dust 2, 7
  noise 98
  radiation 122, 124
Petri-slides 15–17

Index

Phase contrast lighting 22, 24
Photograph 129
Photometer
  corrections 111–12
  Hagner Universal 112–13, 118
  photo-cell 111, 115
  photo-electric 112, 116
Pipette, glass 30
Piston hand pump 36, 45, 49
  checks 49
  stroke 47, 49
Pistonphone 100
Pitot-static tube 76–8, 80–3
Pollens 2
Pollutant
  air 86–9
  collection of 30
  workplace 40, 128
Pollution 70
  workplace 128
Pore size 3
Pressure
  barometric 82–3, 92–3
  instruments for 79
  gauge 71, 76–7, 80–2
  measuring instruments 71
  standard air 82
  static 71, 76–7, 82
  transducer 79
  velocity 71, 76–7
  ventilation 85
Pre-tube 47
Psychrometer
  aspirated 54–5
  sling 53–4, 59, 63
Psychrometric chart 54, 59, 62, 64
  CIBS 61
Psychrometry 52
Pulse counter 17
Pump, 3–7, 12–19, 30
  air 2, 32
  bellows 45–7
  characteristics 6
  controlled flow rate 7, 19
  double acting piston 5, 6, 33
  dry vane rotary 5, 6, 15, 20, 33
  hand 30, 45
  high flow rate 7
  low flow rate 13, 33, 37, 41–2
  manufacturers of 6
  medium flow rate 7, 13
  piston type 45
  reciprocating action 20
  rotor 6
  single acting diaphragm 5–7, 33
  stroke 47
    counter 39, 40, 48
  suction 2, 36, 44
  types 5, 6
Pure tone 94

Quartz crystal 8

Radar transmitters 126
Radiant heat 52–3
Radiation 52, 122–7
  ionising 122
  instruments 123
  intensity 122–4
  non-ionising 122–4
  personal exposure 124
Radioactive material 80, 85
Radio transmitters 126
RAM 12
RDM 11, 12
Reflectance 110, 120
Reflections 117
Respirable
  fibres 24
  mass monitor 10
  materials 29
  particles 5
    samplers 9–11
    size selection 5, 12
Respiratory passages 2
Results sheet
  anemometer calibration 92
  daylight factors 115
  dust sampling 18
  illuminance 119
  lighting data 121
  luminance and reflectance 120
  octave band analysis 104
  rotameter calibration 14
Rotameter
  calibration of 12–17
  float 14–16
  sponge pad 16, 20
Royco 8, 12
Rubber bulb 13

Samplers, pump and tube 33
Sampling
  air 30
  bags 30, 42
  bubbler 43
  duration 42
  grab 29–30
  methods of 2, 31
  passive 32
  period 2, 17, 30–2, 39–41
  personal 2, 3, 7, 30, 32, 37, 43,
  point 32, 42
  position 18
  size selection 5
  static 2–4, 7, 43
  train 2, 6, 8, 11, 16–17, 44
  volume 22
Sand 2
Saturation, percentage 59, 62
Scale factor 73, 85
Scintillation counter 123–4
Sensitising agent 29
Shadows 117
Shelf life 36
Sibata 10, 12

153

Index

Simpson Committee 25
SIMSLIN II 9–12
Skin burns 125
Sling psychrometer 53–4, 59, 63
Smoke tube kit 78, 87
Soap
  bubble 12–15, 34
  liquid 13, 15
Solvent 29
  leaching 35
  paint 40
  vapour 37
Sound
  duration of 94
  frequency 94, 96, 102
    speech 102
  intensity 94–6, 103
  level meter
    grades, types 98, 99, 102
    impulse 97–100
    response of 101
  measuring instruments 94
    pressure 95
    weighting 96
  pitch 94
  spectrum 94, 96, 103, 105
Specific
  enthalpy 59, 62
  gravity 73
  volume 59, 62
Spores 2
Stain
  indicating 45–6
  length of 35, 45–7
  paper tape 37
Standards 129, 130
  asbestos 25
  hygiene 17
  IES Code 117
  microwave 126
  noise exposure 101, 106
Static sampling 2–7, 15, 16, 42
Statistical distribution analyser 97
Stop watch 13, 63, 74
Stress from noise 100–102
Stroboscopic effect 117
Stroke counter 7
Suction inlet 86–9
Sulphur dioxide 30, 37
Sumps 45
Suppliers of equipment 142–3
Survey 18
  check lists 131–141
  lighting 116
  results 129
  thermal 62–6
Syringe 30

Tape recording 97
Temperature 52, 59
  air 52, 92–3
  continuous recording of 55
  dry bulb 52, 57, 59, 62–5
  globe 52, 57, 59, 62
  mean radiant 52, 66–7
  natural wet bulb 53, 63–5
  probe 54
  ventilated wet bulb 66
  WBGT 59, 62, 65
  wet bulb 52, 57, 59, 62–5
Thermal
  environment 52–3, 62
  load 62
  survey check list 135
Thermo-hydrograph 55, 57
Thermoluminescent dosemeter 124
Thermometer 52–5, 63, 66
  alcohol in glass 57
  Botsball 59
  dry bulb 53, 54, 63, 66, 69, 91
  electrical 53
  globe 52, 57–9, 63, 65–6, 69
  Kata 57, 59, 68–9
  mercury in glass 53, 57, 63
  natural wet bulb 64
  resistance 53
  shielded dry bulb 64
  thermo-couple 53
  wet bulb 53–4, 66
Thermos flask 63
Threshold limit value 17, 62
T-piece 13
Tracer gas 87, 89
Tripod stand 2, 15, 17, 44, 63, 100, 101
Toxins 29
TSI Respirable Mass Monitor 10, 12
Tube holders 34, 39
Tubes 32
  absorbent 33–7, 40
    holders for 34, 37
    types of 30–9
  charcoal 30
  colourimetric detector 32
  metal 35
  pitot-static 76–8, 80–3
  pre- 47
  two-stage 41
  U- 71–3
Tubing
  connecting 2, 3
  flexible plastic 13–17, 84–6
Turbulence 80, 87
Tyndall beam 27

U-tube 71–3
Utilisation factor 110–11, 119
Ultrasound 124–5
Ultra-violet radiation 124–5

Vacuum operated personal samplers 30–1
Vapour
  pressure 59
  water 53, 59
Vapours 29–51
Velocity
  air 52, 57, 66, 74–7, 83

Index

contours 88
  measuring instruments 74, 78
  pressure 71
Ventilation 70–93
  flow rate 84
  mechanical 70, 89
  natural 70, 89–91
  pressure 85–6
  record sheet 85–6
  survey check list 137
  systems 70, 79, 85
Visual assessment 117
  results 121
Volume
  flow 14, 17–26, 33–4, 39–45, 70, 83, 89
  specific 59, 62

Water
  distilled 53–4, 63
  vapour 53, 59
WBGT 59, 62, 66
Weight gain 19
Welding, fumes from 26
Whirling hygrometer 53–4
Wick 53, 59, 63, 66
Wind tunnel 79
  open jet 79, 91–3
Work
  rate 62, 65–6
  rest periods 62, 65
  rest regime 65
Work-place environment 62

X-ray diffraction analysis 26
X-rays 122

Yaw, angle of 75–6

155